D1532281

Methods in Molecular Biology

Volume 1

PROTEINS

Biological Methods

Methods in Molecular Biology, edited by **John M. Walker,** *1984*
Volume I: ***Proteins***
Volume II: ***Nucleic Acids***

Liquid Chromatography in Clinical Analysis, edited by **Pokar M. Kabra** and **Laurence J. Marton,** *1981*

Metal Carcinogenesis Testing: *Principles and In Vitro Methods,* by **Max Costa,** *1980*

Methods in Molecular Biology

Volume 1

PROTEINS

Edited by

John M. Walker

Humana Press • Clifton, New Jersey

Library of Congress Cataloging in Publication Data

Main entry under title:

Methods in molecular biology.

Includes bibliographies and index.
Contents: v. 1. Proteins.
1. Molecular biology—Technique—Collected works.
I. Walker, John M., 1948–
QH506.M45 1984 574.8′8′078 84-15696
ISBN 0-89603-062-8

Crescent Manor
PO Box 2148
Clifton, NJ 07015

Printed in the United States of America

Preface

In recent years there has been a tremendous increase in our understanding of the functioning of the cell at the molecular level. This has been achieved in the main by the invention and development of new methodology, particularly in that area generally referred to as "genetic engineering". While this revolution has been taking place in the field of nucleic acids research, the protein chemist has at the same time developed fresh methodology to keep pace with the requirements of present day molecular biology. Today's molecular biologist can no longer be content with being an expert in one particular area alone. He/she needs to be equally competent in the laboratory at handling DNA, RNA, and proteins, moving from one area to another as required by the problem he/she is trying to solve. Although many of the new techniques in molecular biology are relatively easy to master, it is often difficult for a researcher to obtain all the relevant information necessary for setting up and successfully applying a new technique. Information is of course available in the research literature, but this often lacks the depth of description that the new user requires. This requirement for in-depth practical details has become apparent by the considerable demand for places on our Molecular Biology Workshops held at Hatfield each summer. This book is therefore an attempt to provide detailed protocols for many of the basic techniques necessary for working with DNA, RNA, and proteins. This volume gives practical procedures for a wide range of protein techniques. A companion volume (Volume 2) provides coverage for nucleic acids techniques. Each method is described by an author who has regularly used the technique in his or her own laboratory. Not all the techniques described necessarily represent the state-of-the-art. They are, however, dependable methods that achieve the desired result.

Each chapter starts with a description of the basic theory behind the method being described. However, the main aim of this book is to describe the practical steps necessary for carrying out the method successfully. The Methods section therefore contains a detailed step-by-step description of a protocol that will result in the successful execution of the method. The Notes section complements the Methods section by indicating any major problems or faults that can occur with the technique, and any possible modifications or alterations.

This book should be particularly useful to those with no previous experience of a technique, and, as such, should appeal to undergraduates (especially project students), postgraduates, and research workers who wish to try a technique for the first time.

John M. Walker

Contents

Contributors

G. S. BAILEY • Department of Chemistry, University of Essex, Colchester, Essex, England

GRAHAM B. DIVALL • Metropolitan Police Forensic Science Laboratory, London, England

WIM GAASTRA • Department of Microbiology, The Technical University of Denmark, Lyngby, Denmark

KEITH GOODERHAM • MRC Clinical and Population Cytogenetics Unit, Western General Hospital, Crewe Road, Edinburgh, United Kingdom

MAKOTO KIMURA • Max-Planck-Institut für Molekulare Genetik, Abteilung Wittmann, Berlin (Dahlem), West Germany

PER KLEMM • Department of Microbiology, The Technical University of Denmark, Lyngby, Denmark

HARRY R. MATTHEWS • University of California, Department of Biological Chemistry, School of Medicine, Davis, California

E. L. V. MAYES • Department of Protein Chemistry, Imperial Cancer Research Fund, Lincoln's Inn Fields, London, England

JEFFREY W. POLLARD • MRC Human Genetic Disease Research Group, Department of Biochemistry, Queen Elizabeth College, University of London, Campden Hill, London, England

BRYAN J. SMITH • Institute of Cancer Research, Royal Cancer Hospital, Chester Beatty Laboratories, London, England

JOHN M. WALKER • School of Biological and Environmental Sciences, The Hatfield Polytechnic, Hatfield, Hertfordshire, England

JAAP H. WATERBORG • University of California, Department of Biological Chemistry, School of Medicine, Davis, California

BRIGITTE WITTMANN-LIEBOLD • Max-Planck-Institut für Molekulare Genetik, Abteilung Wittmann, Berlin (Dahlem), West Germany

J. N. WOOD • Department of Experimental Immunobiology, Wellcome Research Laboratories, Beckenham, Kent, England

Chapter 1

The Lowry Method for Protein Quantitation

Jaap H. Waterborg and Harry R. Matthews

University of California, Department of Biological Chemistry, School of Medicine, Davis, California

Introduction

The most accurate method of determining protein concentration is probably acid hydrolysis followed by amino acid analysis. Most other methods are sensitive to the amino acid composition of the protein and absolute concentrations cannot be obtained. The procedure of Lowry et al. (*1*) is no exception, but its sensitivity is moderately constant from protein to protein, and it has been so widely used that Lowry protein estimations are a completely acceptable alternative to a rigorous absolute determination in almost all circumstances where protein mixtures or crude extracts are involved.

Materials

1. *Complex-forming reagent*: prepare immediately before use by mixing the following 3 stock solutions A, B, and C in the proportion 100:1:1, respectively.

Solution A: 2% (w/v) Na_2CO_3 in distilled water
Solution B: 1% (w/v) $CuSO_4 \cdot 5H_2O$ in distilled water
Solution C: 2% (w/v) sodium potassium tartrate in distilled water

2. *2N NaOH*
3. *Folin reagent* (commercially available): Use at 1*N* concentration.
4. *Standards*: Use a stock solution of standard protein (e.g., bovine serum albumin fraction V) containing 4 mg/mL protein in distilled water stored frozen at − 20°C. Prepare standards by diluting the stock solution with distilled water as follows:

Stock solution	μL	0	1.25	2.50	6.25	12.5	25.0	62.5	125	250
Water	μL	500	499	498	494	488	475	438	375	250
Protein concentration	μg/mL	0	10	20	50	100	200	500	1000	2000

Method

1. To 0.1 mL of sample or standard, add 0.1 mL of 2*N* NaOH. Hydrolyze at 100°C for 10 min in a heating block or a boiling water bath.
2. Cool the hydrolyzate to room temperature and add 1 mL of freshly mixed complex-forming reagent. Let the solution stand at room temperature for 10 min.
3. Add 0.1 mL of Folin reagent, using a Vortex mixer, and let the mixture stand at room temperature for 30–60 min (do not exceed 60 min).
4. Read the absorbance at 750 nm if the protein concentration was below 500 μg/mL or at 550 nm if the protein concentration was between 100 and 2000 μg/mL.
5. Plot a standard curve of absorbance as a function of initial protein concentration and use it to determine the unknown protein concentrations.

Notes

1. If the sample is available as a precipitate, then dissolve the precipitate in 2*N* NaOH and hydrolyze as in step 1. Carry 0.2 mL aliquots of the hydrolyzate forward to step 2.
2. Whole cells or other complex samples may need pretreatment, as described for the Burton assay for DNA (See Vol. 2). For example, the PCA/ethanol precipitate from extraction I may be used directly for the Lowry assay or the pellets remaining after the PCA hydrolysis (step 3 of the Burton assay) may be used for Lowry. In this latter case, both DNA and protein concentrations may be obtained from the same sample.
3. Rapid mixing as the Folin reagent is added is important for reproducibility.
4. A set of standards is needed with each group of assays, preferably in duplicate. Duplicate or triplicate unknowns are recommended.

References

1. Lowry, O. H., Rosebrough, N. J., Farr, A. L., and Randall, R. J. (1951) Protein measurement with the Folin phenol reagent. *J. Biol. Chem.* **193,** 265–275.

Chapter 2

Determination of Protein Molecular Weights by Gel Permeation High Pressure Liquid Chromatography

E. L. V. Mayes

Department of Protein Chemistry, Imperial Cancer Research Fund, Lincoln's Inn Fields, London, England

Introduction

An essential part of the characterization of any protein is the determination of its molecular weight. The method of choice for these determinations, because of its simplicity and rapidity, has most frequently been sodium dodecyl sulfate (SDS) gel electrophoresis (*See* Chapter 6). The alternative method of gel filtration under denaturing conditions (*1,2*) has not been so widely used, in part as a result of the longer times required for a single run. However, recent developments in gel filtration supports for high pressure liquid chromatography (HPLC) have made more rapid separations possible, and thus gel permeation HPLC is becoming a widely used technique for molecular weight determinations (*3,4*). Gel permeation HPLC, in addition to taking less time than SDS gel electrophoresis, allows easier quantitation and recovery of separated proteins, and the resolution is better than that achieved by gel filtration with conventional materials.

The volume accessible to a protein in gel filtration supports depends on both its size and shape. Thus in order to determine molecular weight the sample protein must have the same shape as the proteins used for calibration. In the presence of denaturants, such as 6*M* guanidine hydrochloride, all proteins in their reduced state adopt a linear random coil conformation whose molecular radius is proportional to molecular weight (*5*). Under these conditions the molecular weight of a protein can be expressed in terms of its elution volume from the column (*1–3*).

Denaturation causes an increase in the intrinsic viscosity of the protein, and hence an increase in its molecular dimensions. Thus, under denaturing conditions the molecular weight exclusion limits of gel filtration matrices are lower than those in the absence of denaturant. Of the column supports used for gel permeation HPLC under denaturing conditions those of the TSK-G3000 SW type are suitable for proteins of less than 70,000 molecular weight (mw), whereas the TSK-G4000 SW type can be used for proteins up to 160,000 mw.

Gel permeation HPLC in guanidine hydrochloride has proved to be a reliable and convenient method for accurate molecular weights determinations (*3,4*). A wide range of sensitivities can be covered, from femtomolar to nanomolar amounts. Because of the high absorbance of guanidine hydrochloride at < 220 nm, the eluate cannot be monitored at this wavelength, therefore proteins in the nanomolar range are detected by absorbance at 280 nm. The sensitivity of detection can be increased by incorporating radioactive label into the protein, for instance by reduction and [^{14}C]-carboxymethylation of cysteine residues.

Gel permeation HPLC is suitable for preparative purposes as well as analytical. Up to approximately 50 nmol can be separated on the standard size columns (7.5 × 600 mm); amounts greater than this reduce the resolution of the separation. Larger amounts can be separated either on preparative columns, or by successive runs with 50 nmol aliquots. For preparative purposes, it may be advantageous to use buffers other than guanidine hydrochloride, such as urea/formic acid (*6*), nondenaturing buffers (*4,7,8*), or volatile buffers (*8,9*). However, the resolution of separation in these buffers is frequently not as good as that in guanidine hydrochloride, and inaccurate estimates of molecular weight will arise from any protein–protein or protein–column interactions that may occur.

Materials

1. 6*M* guanidine hydrochloride (GdHCl) in 0.1*M* potassium dihydrogen phosphate. The pH of the buffer does not require adjustment and should be approximately 4.5. GdHCl of the highest grade available must be used. Filter the buffer through a 0.45 micron filter and thoroughly degas (either by applying a vacuum for 5–10 min until bubbling ceases, or by bubbling helium through the buffer for at least 5 min). Also filter and degas solutions of water and methanol (HPLC grade) for washing the column after use.
2. HPLC equipment, including an absorbance detector for 280 nm and a chart recorder.
3. Column for gel permeation HPLC, e.g., TSK-G type made by Toya Soda and sold by several manufacturers.
4. 0.5*M* Tris HCl, pH 8.5; 6*M* GdHCl
5. 10 m*M* Tris HCl, pH 8.5; 6*M* GdHCl
6. 1*M* solution of dithiothreitol. Store at −20°C in aliquots; after thawing, use immediately and do not refreeze.
7. [^{14}C]-iodoacetamide (40–60 mCi/mmol) dissolved in 0.5*M* Tris HCl, pH 8.5 and 6*M* GdHCl, and then stored at −20°C. Iodoacetamide (nonlabeled), stored at 4°C in the dark (if any yellow color is present, then recrystallize from heptanol).

8. 2,4-Dinitrophenyl (DNP)-lysine and Blue Dextran 2000, approximately 5 mg/mL solutions of each in 6*M* GdHCl, 0.1*M* KH_2PO_4.

Method

1. Reduce and carboxymethylate the protein standards and the sample protein as described in Chapter 5, but dialyze first against 10 m*M* Tris HCl (pH 8.5) and 6*M* GdHCl, then against 6*M* GdHCl in 0.1*M* KH_2PO_4. Suitable protein standards are listed in Table 1.
2. Flush the column at 1 mL/min with water for at least 30 min, followed by 6*M* GdHCl, 0.1*M* KH_2PO_4 until the absorbance at 280 nm drops and becomes stable.
3. Zero the chart recorder and reduce the pump flow rate to 0.5 mL/min.
4. Determine the void volume (V_o) and total available volume (V_t) of the column by injecting a sample containing 90 μL Blue Dextran and 10 μL DNP–lysine solu-

TABLE 1
Protein Standards for Gel Permeation HPLC in Guanidine Hydrochloride

Column type	Protein	Molecular weight
TSK-G3000SW	Bovine serum albumin	66,300
	Ovalbumin	43,000
	α-Chymotrypsinogen	25,700
	Lysosyme	14,400
	Trypsin inhibitor (bovine)	6,500
	Insulin, B chain	3,420
	Insulin, A chain	2,380
TSK-G4000SW	RNA polymerase (*E. coli*) β and β′ subunits	160,000
	Phosphorylase 6	92,500
	Bovine serum albumin	66,300
	Ovalbumin	43,000
	RNA polymerase (E. coli), α subunit	39,000
	Trypsin inhibitor (soybean)	21,000
	Lysosyme	14,400

tions (peaks of a suitable height will be obtained at 280 nm on the 0–0.08 absorbance range). Measure the distance on the chart recorder paper between the sample application and the location of the first and last peaks. V_o and V_t can then be calculated from the distances obtained for the first and last peaks respectively. After the DNP–lysine has been eluted, inject a mix of standard proteins, using amounts equivalent to the sample protein. Adjust the sensitivity range to obtain suitable peak heights.

5. For each protein, calculate the elution volume (V_e) from the distance on the chart recorder between the sample application and the protein peak. The distribution coefficient (K_d) can then be calculated using the following equation:

$$K_d = \left(\frac{V_e - V_o}{V_t - V_o} \right)$$

A plot of molecular weight (M) to the power 0.555 versus K_d to the power 0.333 will give a linear relationship. In Fig. 1 an elution profile and $M^{0.555}$ vs $K_d^{0.333}$ plot are shown for a range of standard proteins on a TSK-G3000 SW.

6. Inject the sample, after all standard proteins have eluted from the column. Calculate K_d and determine the molecular weight of the sample protein from the $M^{0.555}$ vs $K_d^{0.333}$ plot.
7. Do not leave the column and HPLC equipment in GdHCl overnight, but flush with water for at least 30 min at a flow rate of 1 mL/min. If the column is not to be used the following day, then flush through with methanol for 30 min at 1 mL/min.

Notes

1. Reduction and carboxymethylation ensures that the protein adopts a linear random coil conformation in 6M GdHCl, and thus the elution volume will be proportional to molecular weight (and hence $M^{0.555}$ vs $K_d^{0.333}$ will be linear). This process also has the additional advantage of allowing the introduction of

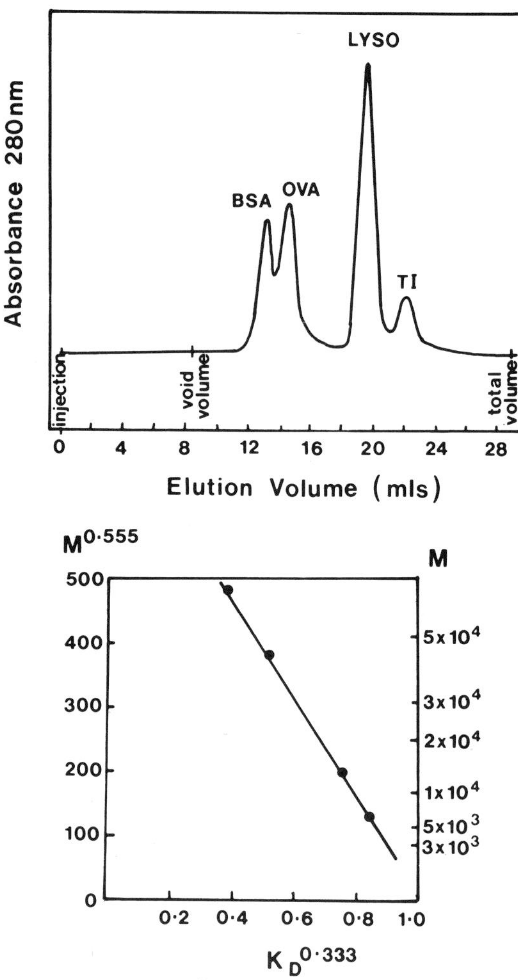

Fig. 1. Typical elution profile obtained from TSK-G3000 SW column in 6*M* guanidine hydrochloride and the calibration curve for molecular weight determinations. (Top) 40 μL of a mixture containing bovine serum albuin (BSA), ovalbumin (OVA), lysosyme (LYSO), and bovine trypsin inhibitor (TI) was injected onto a TSK-G3000 SW column (Pharmacia), equilibrated in 6*M* guanidine hydrochloride, 0.1*M* KH_2PO_4 with a flow rate of 0.5 mL/min. The eluate was monitored at 280 nm. The void volume and total volume of the column were previously determined with Blue Dextran 2000 and 2,4-dinitrophenyl–lysine. (Bottom) The data were plotted according to the equation given in the text to give the calibration curve for molecular weight (*M*) determination.

^{14}C-label into the protein, thus enabling more sensitive detection than absorbance at 280 nm. If ^{14}C-label is not required, the [^{14}C]-iodoacetamide may be omitted from the carboxymethylation procedure.

2. On TSK-G4000 SW columns, Blue Dextran 2000 may give two peaks, one at the void volume and one after the void volume. Thus ensure that V_o is determined from the first peak. For TSK-G3000 SW columns, ^3H_2O gives more accurate determinations of V_t than does DNP–lysine. However, small inaccuracies in either V_o or V_t only effect the slope of the $M^{0.555}$ vs $K_d{}^{0.333}$ plot, and do not effect the estimation of molecular weight.
3. The chloride ions present in GdHCl will, when under high pressure, corrode stainless steel. It is therefore vital that GdHCl is not left in the HPLC equipment for long periods—hence the necessity to flush out the equipment with water each day.
4. To preserve the life of the column, all samples should be centrifuged before injecting into the system. If possible a guard column sho… be attached in front of the analytical column.

References

1. Davison, P. F. (1968) Proteins in denaturing solvents: gel exclusion studies. *Science* **161,** 906–907.
2. Fish, W. W., Mann, K. G., and Tanford, C. (1969) The estimation of polypeptide chain molecular weights by gel filtration in 6*M* guanidine hydrochloride. *J. Biol. Chem.* **244,** 4989–4994.
3. Ui, N. (1979) Rapid estimation of the molecular weights of protein polypeptide chains using high-pressure liquid chromatography in 6*M* guanidine hydrochloride. *Anal. Biochem.* **97,** 65–71.
4. Montelaro, R. C., West, M., and Issel, C. J. (1981) High-performance gel permeation chromatography of protein in denaturing solvents and its application to the analysis of enveloped virus polypeptides. *Anal. Biochem.* **114,** 398–406.
5. Tanford, C. (1968) Protein denaturation. *Advan. Protein Chem.* **23,** 121–282.
6. Waterfield, M. D., and Scrace, G. T. (1981) Peptide separation by liquid chromatography using size exclusion and reverse-phase columns. In *Chromatographic Science Series* (Hawk, G. L., ed.) vol. 18, pp. 135–158, Dekker, New York.

7. Jenik, R. A., and Porter, J. W. (1981) High-performance liquid chromatography of proteins by gel permeation chromatography. *Anal. Biochem.* **111,** 184–188.
8. Lazure, C., Dennis, M., Rochemont, J., Seidah, N. G., and Chretien, M. (1982) Purification of radiolabelled and native polypeptides by gel permeation high-performance liquid chromatography. *Anal. Biochem.* **125,** 406–414.
9. Swergold, G. D., and Rubin, C. S. (1983) High-performance gel-permeation chromatography of polypeptides in a volatile solvent: rapid resolution and molecular weight estimations of proteins and peptides on a column of TSK-G3000-PW. *Anal. Biochem.* **131,** 295–300.

Chapter 3

Immunoaffinity Purification of Protein Antigens

E. L. V. Mayes

Department of Protein Chemistry, Imperial Cancer Research Fund, Lincoln's Inn Fields, London, England

Introduction

The unique high specificity of antibodies, both polyclonal and monoclonal, makes them extremely valuable tools for rapid, selective purification of antigens. In principle, the antibody immobilized on a column support is used to selectively adsorb antigen from a mixture containing many other proteins (*1,2*). The other proteins, for which the antibody has no affinity, may be washed away, and the purified antigen then eluted from the immunoadsorbent. In order to dissociate the antigen from its high affinity antibody, the conditions for elution are necessarily extreme (*1–3*) and thus must be carefully chosen to permit isolation of active protein.

Although the principles involved are relatively simple, in practice many pitfalls may be encountered. However, by adopting a systematic approach to the development and optimization of the purification protocol, these pitfalls can usually be avoided.

Materials

1. Monoclonal antibody, partially purified from ascites fluid or hybridoma supernatant by ammonium sulfate precipitation, ion-exchange chromatography (DEAE-cellulose), or affinity chromatography with either DEAE-Affi-Gel Blue (BioRad) or protein-A Sepharose (see Chapters 28–31).
2. Affi-Gel 10 (BioRad), an *N*-hydroxysuccinimide ester of derivatized crosslinked agarose gel beads.
3. 0.1*M* HEPES, pH 7.5.
4. 1*M* ethanolamine HCl, pH 8.
5. 10 m*M* sodium phosphate, 0.15*M* sodium chloride, pH 7.4 (PBS).
6. 10 m*M* sodium phosphate, 1*M* sodium chloride, pH 7.4 (high salt buffer).
7. 50 m*M* sodium citrate pH 3.
8. 1*M* Tris.
9. Stock solution of sodium azide (20%).

Method

Preparation of Immunoadsorbent

1. Dialyze the antibody solution (0.5–5.0 mg/mL) against HEPES buffer to remove all spurious primary amines such as Tris and free amino acids.
2. Wash the Affi-Gel with 3 bed-volumes of isopropanol, followed by 3 bed-volumes of ice-cold deionized distilled water. This washing procedure must be completed within 20 min and is best performed by filtration in a small Buchner funnel.
3. Add the antibody solution to the washed gel (1 mg antibody/mL of gel) and agitate gently (e.g., by rotating

end-over-end) overnight at 4°C. Do not use magnetic stirrers since these may fragment the gel beads.

4. Add 0.1 mL ethanolamine/mL gel, and agitate for 1 hour further to block all unreacted ester groups.
5. Assay the supernatant for unbound antibody; usually less than 10% of the antibody remains unbound. Since the *N*-hydroxy-succinimide released during the coupling adsorbs at 280 nm above pH 2, dilute an aliquot of the supernatant in 0.1*M* HCl and measure the protein absorbance at 280 nm.
6. Wash the immunoadsorbent with HEPES buffer until all reactants are removed and the absorbance at 280 nm is zero.
7. Precycle the immunoadsorbent with the buffers to be used for the purification procedure by washing with 5 bed-volumes of high salt buffer followed by 5 volumes of sodium citrate buffer.
8. Wash the immunoadsorbent with PBS to return the pH to neutrality and store at 4°C in the presence of 0.02% sodium azide.

Immunoaffinity Purification

1. Incubate the antigen containing extract (in PBS) with the immunoadsorbent by rotating end-over-end for 2 h at 4°C.
2. Remove the supernatant by filtration and wash the immunoadsorbent with at least 10 bed-volumes of high salt buffer followed by 5 volumes of PBS.
3. Elute the antigen batchwise as follows. Gently agitate the immunoadsorbent with 1 bed-volume of sodium citrate buffer for 10 min at 4°C. Remove the eluate by filtration. Repeat with three further applications of sodium citrate buffer. Alternatively the immunoadsorbent may be packed into a column and eluted with sodium citrate buffer at a flow rate of 1 mL/min, or less.
4. Adjust the pH of the eluate to neutrality with 1*M* Tris.
5. Wash the immunoadsorbent with 5 bed-volumes of PBS and store at 4°C in PBS, 0.02% sodium azide.

Notes

1. Since the most favorable conditions for purification may differ for each antigen and antibody, small-scale trials should be made to systematically optimize the various stages of the purification. In order to do this, a suitable assay for the antigen must be available (e.g., by enzymatic activity, radioimmunoassay, or SDS gel electrophoresis). The amounts of antigen and total protein are therefore estimated in the extract before and after immunoadsorption, in the wash buffers and in the eluate.
   The various stages that can be optimized are discussed below.

   (a) Antibody-to-Coupling-Gel Ratio. Usually 1–10 mg of antibody is coupled per mL of gel. High concentrations may lead to reduced efficiency of binding because of steric hindrance, or to such strong binding of the antigen that it cannot be readily eluted.
   (b) Coupling Matrix. Cyanogen bromide (CNBr)-activated Sepharose (available commercially from Pharmacia) can also be used to couple antibody. With both Affi-Gel and CNBr-activated Sepharose, the orientation of the antibody is random, and therefore the efficiency of the antigen–antibody interaction may be decreased. To overcome this, the antibody may be coupled using chemical crosslinkers, such as dimethyl pimelimidate, to either protein A-Sepharose or anti-immunoglobulin crosslinked to Affi-Gel or CNBr-activated Sepharose.
   (c) Optimum Conditions for Antigen Binding. Buffers other than PBS may be more suitable for particular antigen purifications; optimum conditions are usually within the pH range 6–9 and in the presence of approximately 0.15*M* salt. The time of incubation of the antigen with immunoadsorbent may also require optimization.
   (d) Reduction of Nonspecific Binding. Nonspecific binding of proteins to the immunoadsorbent may be minimized by addition to the antigen-containing solution of a nonantigen protein (e.g.,

bovine serum albumin), a non-ionic detergent (e.g., Triton X-100), or an organic solvent (e.g., ethylene glycol). Altering the pH or increasing the ionic strength of the antigen solution may also decrease nonspecific binding. Alternatively, a nonimmune antibody matrix (coupled in the same manner as the immunoadsorbent) can be used to pre-adsorb the extract before incubation with the immunoaffinity matrix.

(e) Antigen-to-Antibody Ratio. Optimal elution and minimal nonspecific adsorption are usually obtained if the coupled antibody is fully saturated with antigen.

(f) Washing Procedures. To remove nonspecifically adsorbed protein, the immunadsorbent should be washed with several column volumes of buffer. Optimal removal is attained by altering either the ionic strength (e.g., 1*M* salt, or distilled water), or pH or by inclusion of organic solvent (e.g., ethylene glycol). Obviously under the conditions chosen the antigen must not be eluted from the immunoadsorbent.

(g) Elution Conditions. The conditions used to elute immunoadsorbents are extreme (Table 1) and may cause denaturation and loss of activity of the antigen. Thus a study of the stability of the antigen under various potential eluting conditions must be made to determine the minimum and maximum conditions that the protein activity will withstand. These conditions can then be tested for elution of the antigen from the immunoadsorbent. Combinations, such as denaturants at low pH, may be required to disrupt antigen–antibody interactions of very high affinity. Specific elution may be possible for some immunoadsorbents, e.g., if the immunodeterminant is carbohydrate, the antigen may be eluted with the carbohydrate or an analogue.

(h) Batch Procedure vs Column. The procedure described in the previous section uses a batch technique throughout; alternatively, the immunoadsorbent may be poured into a column at any stage during the purification. When the antigen extract is applied to the immunoadsorbent packed in a col-

TABLE 1
Conditions for Elution from Immunoadsorbents

Low pH (<3)	50 m*M* glycine HCl, pH 2–3 50 m*M* sodium citrate, pH 2.5–3 1*M* propionic acid
High pH (<10)	0.15*M* ammonium hydroxide, pH 11 0.15*M* NaCl 50 m*M* diethylamine, pH 11.5 0.5% sodium deoxycholate
Denaturants	4*M* guanidine hydrochloride 8*M* urea
Chaotropic ions (up to 3 *M*)	3*M* sodium isothiocyanate
($CCl_3COO^- >> SCN^- > CF_3COO^- > ClO_4 > I^- > Cl^-$)	3*M* sodium iodide
Polarity reducing	Ethylene glycol (50%) Dioxan (10%)
Low ionic strength	Distilled water

umn, the flow rate should be low, < 20 mL/h. Recycling the flow-through may allow further binding of antigen. The direction of flow should be reversed for elution to prevent strongly bound antigens from interacting with the immunoadsorbent throughout the column before elution. If the affinity of the antibody for antigen is high, elution may be helped by stopping the flow for 10 min or longer when the column has been equilibrated in elution buffer.

2. Monospecific polyclonal antiserum, obtained by immunization with purified protein, may be used instead of monoclonal antibody. The serum should be purified before use by one of the techniques suggested for monoclonal antibody purification. (*See* Chapter 31).
3. Immunoaffinity chromatography has been successfully used to purify several membrane proteins. Because of the hydrophobic nature of these proteins, all solutions used throughout the purification must contain deter-

gent (e.g., Triton X-100, NP40, or sodium deoxycholate). Sodium deoxycholate is extremely effective for solubilizing membrane proteins; however, it is incompatible with buffers of high ionic strength, or pH of less than 8.0 since a viscous gel is formed. One detergent may be exchanged for another during washing while the antigen is bound to the immunoadsorbent. Detergents may reduce the affinity of antibody for antigen; thus, pilot studies must be made to determine the optimum conditions for solubilization and the antigen–antibody interaction.

4. If the antigen does not bind to the immunoadsorbent despite optimization of the extract conditions, then either the affinity of the antibody is too low or the coupling procedure has caused denaturation of the antibody. The former cause is intrinsic to the antibody and cannot be remedied. Denaturation upon coupling may be circumvented by use of a different coupling support, or by partially blocking some of the coupling groups prior to addition of antibody (thus the gel is first incubated with coupling buffer alone for a few hours, and then with the antibody). Alternatively the antibody may be noncovalently bound to immobilized protein A or anti-immunoglobulin. With this type of immunoadsorbent, antibody and antigen will be eluted together and therefore a further purification step will be required.
5. Another problem that may be encountered is that the antigen sometimes cannot be eluted from the immunoadsorbent. This may be the result of a very high affinity of the antibody for antigen. Since Fab fragments frequently have lower affinities than the antibodies from which they are derived, an immunoadsorbent made with Fab fragments may allow elution of the antigen. Alternatively the antigen may elute from the immunoadsorbent, but in an inactive form and therefore would not be detected. To test whether this is the case, SDS gel electrophoresis should be used to analyze both the eluate, and an aliquot of the immunoadsorbent (boiled in 2% SDS for 2 min) for protein of the appropriate molecular weight.
6. The life of the immunoadsorbent will depend on how frequently it is used and the conditions required to

elute the antigen. The immunoadsorbent should be returned to normal conditions as soon as possible after elution, and should be stored at 4°C (not frozen) in the presence of a bacterial inhibitor. Centrifuging the antigen-containing extract to remove particulate matter will also preserve the life of the immunoadsorbent. Also if proteases are present in the extract, inhibitors such as phenylmethyl sulfonyl fluoride (2 m*M*), iodoacetamide (2.5 m*M*), and EDTA (5 m*M*), should be added to protect both the antigen and antibody from proteolysis.

References

1. Livingston, D. M. (1974) Immunoaffinity Chromatography of Proteins in *Methods in Enzymology* (eds. W. B. Jakoby and M. Wilchek) vol. 34, pp. 723–731, Academic Press, New York.
2. Dalchau, R., and Fabre, J. W. (1982) The Purification of Antigens and Other Studies with Monoclonal Antibody Affinity Columns: the Complimentary New Dimension of Monoclonal Antibodies in *Monoclonal Antibodies in Clinical Medicine* (eds. A. J. McMichael and J. W. Fabre) pp. 519–556, Academic Press, New York.
3. *Affinity Chromatography. Principles and Methods.* Technical brochure available from Pharmacia Fine Chemicals, Sweden.

Chapter 4

Electrophoretic and Chromatographic Separation of Peptides on Paper

E. L. V. Mayes

Department of Protein Chemistry, Imperial Cancer Research Fund, Lincoln's Inn Fields, London, England

Introduction

Electrophoresis and chromatography on paper have proved extremely useful techniques for the separation of small peptides prior to amino acid sequencing. Polarity determines the rate of migration of a peptide during chromatography, whereas both charge and size are the main determinants during electrophoresis. The two techniques therefore are complimentary, and when used in conjunction can resolve most peptide mixtures into their pure components.

Ideally the mixture to be separated should contain less than 20 peptides, since only a limited amount of a mixture can be loaded per chromatogram. These techniques therefore are frequently used for final separation following an initial purification by either gel filtration or ion-exchange chromatography. On Whatman 3 MM paper up to 0.5 mg of peptide mixture can be loaded per centimeter (fivefold less on Whatman No. 1); heavier loadings result in distortion and streaking of the peptide bands with a concomitant decrease in resolution. Optimum resolution is achieved with peptides of less than 20 amino acids, thus these techniques are usually used for separation of peptides obtained by enzymatic digestion (e.g., by trypsin, or *Staphylococcus aureus* V8 protease) rather than those obtained by chemical cleavage (e.g., by cyanogen bromide). Peptides larger than 20 amino acids remain at the origin in both electrophoresis and chromatography, and in addition their presence can cause streaking of smaller peptides during electrophoresis. The peptide mixture must be free from salts that cause severe streaking of the peptide bands; thus the initial purification steps, or the enzymatic digestion are best carried out in volatile buffers (e.g., ammonium bicarbonate, acetic acid, pyridine acetate).

Several stains can be used to visualize the peptides; some react with amino groups and therefore stain all peptides, e.g., ninhydrin, fluorescamine (*1*), whereas others are specific for peptides containing a particular amino acid, e.g., phenanthrenequinone for arginine (*2*) or the Ehrlich test for tryptophan. Some stains are compatible with others, thus allowing one chromatogram to be stained both for specific amino acids, and for all peptides (*3*). All these stains are destructive, and make the peptides unsuitable for amino acid sequencing. Thus for preparative purposes narrow strips from the edges of the chromatogram are stained and the corresponding peptides marked and eluted from the unstained portion. Alternatively the whole chromatogram is stained with a weak solution of fluorescamine (*1*), which reacts with only a small percentage of each peptide, but is sufficiently sensitive to allow their detection.

Trial runs with small aliquots of the peptide mixture are carried out in order to determine the optimum conditions for separation. By judicious choice of the solvent mixture and the duration of the separation, all the peptides in a mixture can frequently be resolved by either chromatography or electrophoresis. More complex mixtures will require separation by one technique followed by the other.

Materials

1. Whatman chromatography paper No. 1 and 3 MM, sheet size 46 × 57 cm.
2. Disposable plastic gloves. These must always be worn when handling the paper, since fingerprints will be stained by the sensitive stains used.
3. Glacial acetic acid, pyridine and *n*-butanol, all A.R. grade.
4. Glass chromatography tanks for descending chromatography (at least 51 × 20 × 56 cm) complete with troughs. High-voltage flat-bed electrophoresis apparatus fitted with safety cutoffs and cooling plates to take a maximum sheet size of 61 × 28 cm, and a high voltage power supply, to give at least 200 mA at 3 kV.
5. Dressmaking pinking shears, sewing machine, and cotton.
6. Electric fan heater or hair dryer.
7. Microsyringe with blunt-ended needle. The most useful sizes are 10 and 50 μL.
8. Ninhydrin spray. Aerosols of ninhydrin solution are commercially available; otherwise a solution (0.25% in acetone) may be prepared and used with a universal aerosol spray.
9. Oven (110°C) for developing ninhydrin-sprayed chromatograms.
10. Flat-bottomed specimen tubes with approximately 10 mL capacity. These are for collecting eluted peptides and should be siliconized with Repelcote.

N.B.: *n*-butanol, pyridine, ninhydrin solution, and Repelcote are all toxic and should therefore be used in a fume hood.

Method

Paper Chromatography

1. Place a folded sheet of Whatman 3 MM in the bottom of the chromatography tank and saturate this with chromatography solvent (*see* Table 1). Fill the trough with

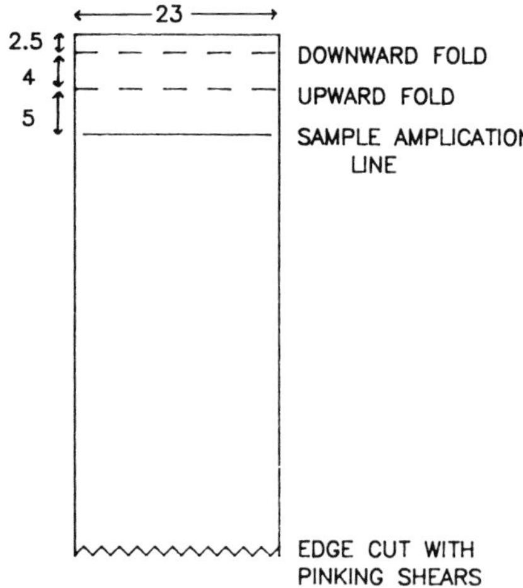

Fig. 1. (a) Folding and marking of paper for chromatography.

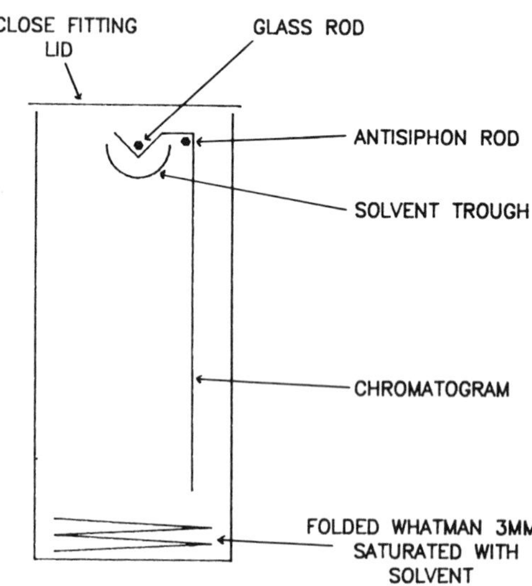

Fig. 1. (b) Positioning of paper in chromatography tank.

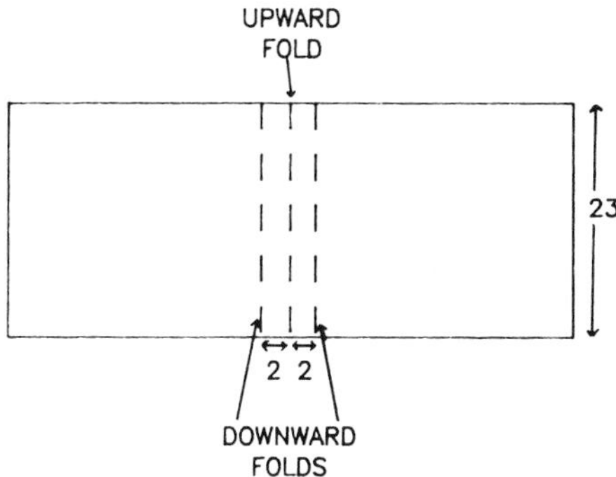

Fig. 1. (c) Folding of paper for electrophoresis. All dimensions are given in centimeters.

solvent, close the lid, and allow the atmosphere within the tank to equilibrate. Meanwhile, prepare the chromatogram.

2. Cut a sheet of Whatman 3 MM in half lengthwise and fold as shown in Fig. 1a. Draw a pencil line 2 cm from the second fold to within 1 cm of either edge—this becomes the guideline for loading the sample. The bottom edge of the sheet should be cut with pinking shears to ensure that the solvent runs evenly across the width of the chromatogram.
3. Dissolve the peptide mixture in aqueous solution (e.g., water, 0.1*M* acetic acid, or 0.1*M* ammonium bicarbonate) and centrifuge to remove any insoluble material. Using a microsyringe, load the sample (0.1–0.5 mg/cm) on the pencil line. The area of application should not touch the bench; the folds are usually sufficient to ensure this does not happen. Do not allow the width of the application strip to exceed 0.3 cm; this may require repeated applications, drying between each with a warm air stream from a fan heater or hair dryer.
4. When the sample is dry, place the chromatogram in the tank with the downward fold in the trough and held in place with a glass rod; the upward fold should be over the antisiphon rod (see Fig. 1b).

5. Leave the chromatogram to develop until optimum separation is achieved. The time required, usually between 1 and 5 d, should be determined previously by trial separations of aliquots of the mixture (see notes).

Paper Electrophoresis

1. Cut a sheet of Whatman 3 MM in half lengthwise and fold as shown in Fig. 1c. The position of the upward fold should previously be determined by trial runs on aliquots of the mixture (see notes), and for most peptide mixtures will be nearer to the anode.
2. Load the peptide mixture along the upward fold as described for paper chromatography.
3. When the sample is dry, evenly wet the electrophoretogram with the appropriate buffer (*see* Table 1). This is best done using a 10 mL pipet. First, run the pipet along the two downward folds, allowing a small amount of solvent to absorb onto the paper, which will then wet the sample application line by capillary action (do not wet the application line directly). The remainder of the paper is then wetted by passing the pipet backwards and forwards across the width of the paper while allowing the buffer to slowly flow out of the pipet (do not overwet). Blot the paper (excluding the sample application line) between two sheets of Whatman 3 MM and place it in the electrophoresis apparatus.
4. Fill the electrode troughs with buffer and make wicks the same width as the electrophoretogram by folding Whatman 3 MM paper. Wet the wicks with buffer, and blot with Whatman 3 MM. Place the wicks in the electrode troughs and overlap them onto the electrophoretogram by approximately 1 cm.

N.B.: Excessive wetting of either the electrophoretogram or wicks leads to condensation, and subsequent short circuiting of the current.

5. Ensure that the water for the cooling plates is turned on, and turn on the power supply. Electrophoresis at 1–2 kV for 1–2 h is usually sufficient for the separation of most peptide mixtures; the exact conditions should be determined previously by trial runs.

Detection of Peptides and Their Elution

1. When the chromatographic or electrophoretic separations have finished, hang the paper to dry in a warm air stream from a fan heater or hair-dryer.
2. Cut 2-cm-wide strips from each long edge of the chromatogram and mark each so that they can later be lined up with the remainder of the sheet.
3. Attach these strips to a sheet of Whatman No. 1 with paper clips, thus keeping them straight and allowing several side strips to be stained simultaneously. Spray the strips evenly with ninhydrin solution (do not overwet since this may cause the peptide bands to streak). After allowing the strips to dry for a few minutes, place them in a 110°C oven for 5 min. The peptides should be clearly visible as blue bands; if not, the spraying and heating process may be repeated.
4. Match the strips with the chromatogram and draw lines across to connect the corresponding peptide bands on the two strips.
5. The peptides are eluted from the unstained paper by descending chromatography with 0.1*M* acetic acid. To do this, cut a sheet of Whatman No. 1 in quarters lengthwise. Sew a quarter sheet along each edge of the chromatogram and cut across the full width for each peptide band. One piece of the Whatman No. 1 is folded as given for chromatography and the other cut into a point (if the peptide band is wider than 3 cm, it is advisable to cut two points). These strips are placed in a chromatography tank with 0.1*M* acetic acid in the trough. Siliconized tubes are placed under each point to collect the eluted peptides (if a large tank is used, the tubes can be placed on a glass shelf supported at the correct height in the tank). Leave for 24 h before removing the tubes and drying the eluted samples *in vacuo* over sodium hydroxide and phosphorus pentoxide.

Notes

1. The optimum conditions for peptide separation within one chromatogram or electrophoretogram are deter-

mined by preliminary trial separations with aliquots of the mixture. Approximately 0.1–0.5 mg of mixture is loaded across a 1 cm strip; thus one-half sheet of Whatman 3 MM can be used for 11 trial separations, each mixture being separated from the next by 1 cm. For chromatography it is only necessary to determine the optimum time for separations, while for elecrophoresis the time, voltage, and position for sample application must be determined. Some commonly used chromatography solvents and electrophoresis buffers are shown in Table 1.

2. Other stains besides ninhydrin can be used for the detection of peptides (*see* Table 2 and Chapter 21). Unfortunately the use of stained side strips does not detect irregularities in the peptide bands across the width of

TABLE 1
Useful Solvents for Paper Chromatography and Electrophoresis

Method	Composition, v/v	
Chromatography	*n*-Butanol : acetic acid : water : pyridine 30 : 9 : 24 : 20	
	n-Butanol : acetic acid : water : pyridine 15 : 3 : 12 : 10	
	n-Butanol : acetic acid : water 4 : 1 : 5	(upper phase only, lower phase in bottom of tank)
Electrophoresis	Pyridine : acetic acid : water 1 : 10 : 289	pH 3.5
	Pyridine : acetic acid : water 3 : 3 : 395	pH 4.8
	Pyridine : glacial acetic acid : water 25 : 1 : 225	pH 6.5
	Formic acid : acetic acid : water 52 : 29 : 919	pH 1.9

the chromatogram. This may be overcome by spraying the chromatogram lightly with a weak solution of fluorescamine (0.001–0.0005% in acetone), after an initial treatment with 3% pyridine in acetone (*1*). Only a small portion of the free amino groups react under these conditions, allowing visualization of the peptides by fluorescence under UV light (366 nm) without rendering all the peptides unsuitable for sequencing. Fluorescamine (0.01% in acetone) is also considerably more sensitive than ninhydrin, detecting as little as 0.1 nmol/cm^2, compared to 5 nmol/cm^2, and thus can also be used for staining side strips.

3. The main problem encountered during paper electrophoresis or chromatography is uneven migration and/or streaking of the peptide bands. Several factors can cause the occurrence of this phenomenon: (a) The presence of nonvolatile salts in the sample; (b) peptides too large (ideally < 20 amino acids); (c) too much sample applied (not more than 0.5 mg/cm); (d) excessive wetness during electrophoresis; (e) insufficient cooling during electrophoresis; (f) an uneven temperature gradient in the chromatography tank (e.g., because of exposure to draft or to direct sunlight); (g) excessive wetting with staining agent.
4. It is essential that pure solvents are used since impurities such as aldehydes may lead to loss of aromatic acids. The addition of a trace of β-mercaptoethanol to the solvent prevents autoxidation of methionine and cysteine derivatives.
5. Cleanliness is essential to avoid contamination of the peptides with either proteases or microorganisms. Thus all surfaces should be cleaned regularly with 70% ethanol to remove microorganisms and dust. Chromatograms must be handled at all times with either disposable plastic gloves or forceps.
6. The purity of eluted peptides can be assessed by trial runs under different conditions as outlined previously. Ultimately *N*-terminal analysis (*see* Chapter 23) indicates whether the peptide is sufficiently pure for amino acid sequencing.
7. Yields from paper depend on the properties of the peptide, and are usually between 60 and 80%, but some-

TABLE 2
Alternative Stains for the Detection of Peptides[a]

Stain	Specificity	Color	Sensitivity
Ninhydrin–collidine (4)			
600 mL ethanol + 200 mL glacial acetic acid + 80 mL collidine + 1 g ninhydrin	Free amino groups	Various (pink, yellow, green, blue purple)	5 nmol cm^{-2}
Store 4°C			
Develop 5–10 min 80°C		Colors fade, store stained papers −20°C	
Iodine–starch(5)			
Spray with iodine solution (0.5% in chloroform)	Peptide bond	Blue black	>5 nmol cm^{-2}
Leave room temperature to dry			
Spray with starch solution (1% in water)			
Ehrlich reagent (3,4)			
2% (w/v) *p*-dimethylaminobenzaldehyde in acetone; immediately before spraying mix 9 vol with 1 vol conc. HCl	Tryptophan	Blue purple	5 nmol cm^{-2}
Develop for a few minutes room temperature			
Can be used after ninhydrin, but will bleach ninhydrin colors			
Phenanthrene-1,2-quinone (2)			
0.02% Phenanthrene-1,2-quinone in absolute ethanol (A), store in dark 4°C for months. 10% (w/v) NaOH in 60% (v/v) aqueous ethanol, (B), store 1 month	Arginine	Greenish-white fluorescence against dark background	0.1 nmol cm^{-2}

Mix equal parts A and B prior to spraying Dry at room temperature, 20 min Visualize with UV (254 or 366 nm)			
Pauly test (3,4)			
1 g sulfanilamide + 10mL 12*M* HCl + 90 mL water, (A); 5 g $NaNO_2$ in 100 mL water, (B); 50 mL saturated Na_2CO_3 + 50 mL water, (C)	Histidine	Cherry red	5 nmol cm^{-2}
Store all solutions 4°C. Mix 5 mL A + 5 mL B	Tyrosine	Dull brown	
Add 40 mL *n*-butanol after 1 min. Shake 1 min, leave to settle, use butanol layer to spray or dip chromatogram. 5 min at room temperature. Spray with C for imidizoles other than histidine			
Nitrosonaphthol Test (3)			
0.1% (w/v) α-nitroso-β-naphthol in acetone (A); acetone:conc HNO_3 (9:1, v/v) (B), freshly prepared	Tyrosine	Rose on yellow; background color fades	10 nmol cm^{-2}
Dip or spray with A. Dry Dip or spray with B. Dry 5–10 min Heat gently with hot air			

[a]*See also* Chapter 21.

times as low as 20%. If smaller amounts of peptides are fractionated Whatman No. 1 may be used and up to 0.1 mg/cm loaded (when wet, Whatman No. 1 tears easily and therefore should be handled with care).

References

1. Vandekerckhove, J., and Van Montagu, M. (1974) Sequence analysis of fluorescamine-stained peptides and proteins purified on a nanomole scale. *Eur. J. Biochem.* **44,** 279–288.
2. Yamada, S., and Itano, H. A. (1966) Phenanthrenequinone as an analytical reagent for arginine and other monosubstituted guanidines. *Biochim. Biophys. Acta* **130,** 538–540.
3. Easley, C. W. (1965) Combinations of specific color reactions useful in the peptide mapping technique. *Biochim. Biphys. Acta* **107,** 386–388.
4. Bennett, J. C. (1967) Paper, chromatography and electrophoresis; special procedure for peptide maps. In *Methods in Enzymology* (ed. Hirs, C. H. W.) vol. XI, pp. 330–339. Academic Press, London.
5. Barrett, G. C. (1962) Iodine as a "Non-Destructive" color reagent in paper and thin layer chromatography. *Nature* **194,** 1171–1172.

Chapter 5

Peptide Mapping by Reverse-Phase High Pressure Liquid Chromatography

E. L. V. Mayes

Department of Protein Chemistry, Imperial Cancer Research Fund, Lincoln's Inn Fields, London, England

Introduction

Reverse-phase high pressure liquid chromatography (HPLC) has proved to be an extremely versatile technique for rapid separation of peptides (reviewed in ref. *1*). One of its uses is for peptide mapping, or "fingerprinting" (*2–5*) as an alternative procedure to the conventional two-dimensional separations on paper or thin-layer supports (*See* Chapter 21). Although the map obtained is one dimensional, the excellent resolving power of reverse-phase HPLC enables separation of the majority of peptides within a mixture. HPLC offers the advantages of high reproducibility, easy quantitation, and rapid analysis time, and is also suitable for automation.

A wide range of sensitivities (from subfemto- to nanomolar amounts) can be covered by use of an appropriate method for detection of the peptides. For nanomolar quantities the eluate is monitored at two or more wavelengths, thus detecting all peptides (at < 220 nm), and only those containing aromatic amino acids (tyrosine and tryptophan at 280 nm; phenylalanine at 255 nm). The sensitivity of detection is increased to picomolar levels by either fluorescence detection of tyrosine and tryptophan (*6*) or postcolumn fluorescent derivatization with fluorescamine (*3,7*) or *o*-phthalaldehyde (*6,7*). By incorporation of radioactive label into proteins, as little as subfemtomolar amounts may be analyzed. To minimize peptide losses at these low levels, unlabeled carrier protein (e.g., bovine serum albumin) should be added prior to proteolysis. Proteins can be labeled biosynthetically with amino acids (e.g., ^{35}S-methionine, ^{35}S-cysteine, ^{3}H-lysine) thus allowing positive identification of precursors produced during in vivo pulse labeling or in vitro translation (*4*). Alternatively, the label can be introduced postsynthetically by modification with radioactive reagents (e.g., ^{125}I-iodination, ^{14}C-carboxymethylation). Radioactive labeling can also be useful for peptide maps on larger amounts of protein, thus allowing the identification of peptides containing specific amino acids (e.g., cysteine or methionine).

For reverse-phase HPLC of peptides, the most suitable column supports are the cyanopropyl-(CN-), ocytl-(C_8-), or octadecyl-(C_{18}-) silica types (the latter two usually give better separation than the CN-silica types). Several buffers have been used for peptide separation, such as orthophosphoric acid (*2*), pyridine–acetic acid (*3*), triethylammonium phosphate (*7*), and trifluoroacetic acid (TFA) (*8*). Peptides are eluted with increasing concentrations of organic solvents, most frequently acetonitrile or *n*-propanol. In this laboratory, TFA/acetonitrile is routinely used, since it is both volatile, and allows high sensitivity detection by absorbance at < 220 nm.

Materials

1. 0.1% TFA (buffer A) and 0.1% TFA/60% acetonitrile (buffer B). TFA and acetonitrile should be HPLC or

sequencing grade, and the water used must be free of organic contaminants (most laboratory distilled water requires further purification, e.g., by use of a Millipore Milli-Q system). Filter the buffers through a 0.45 μm filter to remove all particulate matter and degas (either by applying a vacuum for 5–10 min until all bubbling ceases, or by bubbling helium through for at least 5 min).

2. HPLC equipment, including a solvent programmer and two detectors, one at 280 nm to monitor tryptophan- and tyrosine-containing peptides and the other within the range 205–215 nm to monitor all peptides.
3. Column for reverse-phase HPLC; either octyl-(C_8-) or octadecyl-(C_{18}-) silica type.
4. 0.5*M* Tris HCl, pH 8.5–6*M* guanidine hydrochloride (GdHCl). The highest grade of GdHCl should be used.
5. 10 m*M* Tris HC1, pH 8.5–2*M* GdHCl.
6. 0.1*M* ammonium bicarbonate.
7. Stock solution of *N*-tosyl-L-phenylalanyl chloromethyl ketone- (TPCK-) treated trypsin, 10 mg/mL in 0.1 m*M* HCl. This solution can be stored at −20°C for months without loss of activity.
8. 1*M* stock solution of dithiothreitol (DTT). Store at −20°C in aliquots; after thawing, use immediately and do not refreeze.
9. [^{14}C]-iodoacetamide (40–60 mCi/mmol) dissolved in 0.5*M* Tris HCl, pH 8.5, 6*M* GdHCl and stored at −20°C. Iodoacetamide (nonlabeled), if any yellow color is present the iodoacetamide must be recrystallized from heptanol. Store the solid at 4°C in the dark.

Methods

Full Reduction and Carboxymethylation

1. Dissolve the protein in 0.5*M* Tris HCl, pH 8.5–6*M* GdHCl at 1–10 mg/mL.
2. Make 10 m*M* in DTT and incubate at 37°C for at least 2 h.
3. Cool on ice and add 50 μCi [^{14}C]-iodoacetamide. Incubate in the dark for 30 min on ice.

4. Add iodoacetamide to 20 m*M* final concentration and incubate for a further 2 h in the dark and on ice.
5. Dialyze against 10 m*M* Tris HCl, pH 8.5–2*M* GdHCl for at least 3 h and then twice against 0.1*M* ammonium bicarbonate.

These dialysis steps remove both the unreacted iodoacetamide (to ensure that methionine residues are not modified) and the GdHCl prior to proteolysis.

Trypsin Digestion

1. Add TPCK-treated trypsin (2% by weight of the protein) and incubate for 24 h at 37°C.
2. Stop the digestion either by injecting onto the HPLC column immediately, or by adding soyabean trypsin inhibitor or *N*-tosyl-L-lysyl chloromethyl ketone (TLCK) in a slight molar excess over the protease.

Reverse-Phase HPLC

1. Set a flow rate of 1 mL/min through the column and program a gradient of 0–100% buffer B over 60 min.
2. Start the gradient and monitor at both wavelengths on the highest sensitivity. If peaks are observed during the run then repeat this process until they disappear (see Notes).
3. To check the system inject a tryptic digest of a standard protein e.g., cytochrome c or bovine serum albumin. The injection volume can be up to 2 mL, but a better resolution is obtained with less than 200 μL. Up to approximately 50 nmol of protein can be injected per run on a 25 cm × 4.6 mm column. Ideally the amount of standard protein should correspond approximately to that of the protein of interest so that the sensitivities of the two detectors can be adjusted to give the optimum peak height.
4. At 5 min after the injection, start the gradient. When the gradient has been completed, maintain 100% buffer B for a further 10 min. The majority of peptides will elute between 0 and 40% acetonitrile (i.e., 0 and 67% buffer B), and should be sharp discrete peaks (see Fig. 1).

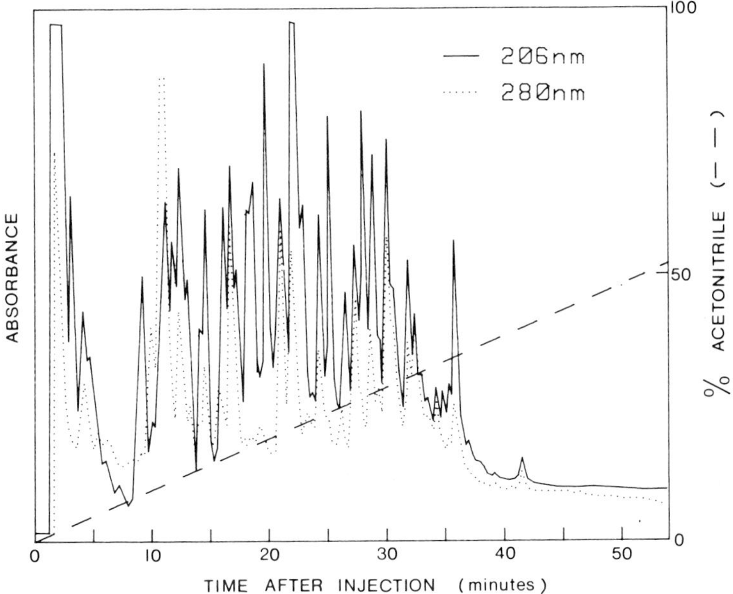

Fig. 1. Tryptic peptide map of bovine serum albumin. 500 μg of bovine serum albumin was digested for 24 h at 37°C with TPCK-trypsin (10 μg). The resultant digest was analyzed on a SynChropak C-18 column (SynChrom Inc.) in 0.1% TFA with an acetonitrile gradient of 0–60% in 60 min. The eluate was monitored at 280 and 206 nm.

5. Allow the column to re-equilibrate at 0% buffer B for 10 min prior to injection of the next digest.
6. Repeat steps 3–5 for each protein digest.

Notes

1. Reduction of disulfide bonds followed by carboxymethylation of cysteine residues prevents the formation of intermolecular disulfide bonds either before or after proteolysis. Under the alkaline conditions used, the iodoacetamide reacts most rapidly with the cysteine residues, then with the thiol groups of the DTT; thus, with a molar ratio of 1:2 of DTT to iodoacetamide, alkylation of methionine residues does not occur. The reaction may be optimized if the moles

of disulfide bonds in the protein are known; use a 2 m*M* excess of DTT over disulfide bonds, and a 1.1-fold molar excess of iodoacetamide over the total thiol groups in the solution.

2. If the protein is insoluble in ammonium bicarbonate, try the addition of urea to 2*M* to aid solubilization. Urea will not effect the digestion, but when injected onto the HPLC column will give a large UV absorbing peak at the void volume; therefore, allow the absorbance to return to baseline level before starting the acetonitrile gradient.
3. At high sensitivities an increase in the absorbance at the lower wavelength may be observed as the proportion of buffer B increases. This results from absorbance by acetonitrile, and may be counteracted by addition of more TFA to buffer A (0.01–0.03%). Peaks may also be observed during the blank run that remain even after several acetonitrile gradients have been passed through the column. These peaks are caused by impurities in the buffers. Provided the peaks are not too large, they can be subtracted from the peptide maps obtained; alternatively, the buffers will require further purification.
4. To preserve the life of the column, all samples must be centrifuged before injection. Also, if possible, a guard column (packed with a similar support material to that of the main column) should be used.
5. Poorly resolved and broad peaks are normally a result of overloading the column with peptides; alternatively, the column may need to be replaced.
6. Peptide mapping by HPLC covers a large range of sensitivity, from very small amounts of radioactively labeled protein up to approximately 50 nmol. Protein may be radioactively labeled in vivo (e.g., ^{35}S-methionine, ^{3}H-lysine) or in vitro (e.g., ^{125}I-iodination or ^{32}P-phosphorylation). When the amount of radioactively labeled protein is small (< 5 μg), a cold carrier protein (e.g., bovine serum albumin) should be added to minimize losses; this has the additional advantage of acting as an internal control on the reproducibility of the separation.

7. In the procedure described here, trypsin is used to digest the protein. This method of peptide mapping is also suitable for use with other proteases that generate small peptides, e.g., *Staphylococcus aureus* V8 protease.

References

1. Hughes, G. J., and Wilson, K. J. (1983) High-performance liquid chromatography: analytical and preparative applications in protein-structure determination, in *Methods of Biochemical Analysis* (ed. Glick, D.) vol. 29, pp. 59–135, Wiley, New York.
2. Fullmer, C. S., and Wasserman, R. H. (1979) Analytical peptide mapping by high performance liquid chromatography. *J. Biol. Chem.* **254,** 7208–7212.
3. Rubinstein, M., Chen-Kiang, S., Stein, S., and Udenfriend, S. (1979) Characterization of proteins and peptides by high-performance liquid chromatography and fluorescence monitoring of their tryptic digests. *Anal. Biochem.* **95,** 117–121.
4. Abercrombie, D. M., Hough, C. J., Seeman, J. R., Brownstein, M. J., Gainer, H., Russell, J. T. and Chaiken, I. M. (1982) Use of reverse-phase high-performance liquid chromatography in structural studies of neurophysins, photolabelled derivatives, and biosynthetic precursors. *Anal. Biochem.* **125,** 395–405.
5. Oray, B., Jahani, M., and Gracy, R. W. (1982) High-sensitivity peptide mapping of triosephosphate isomerase: a comparison of high-performance liquid chromatography with two-dimensional thin-layer methods. *Anal. Biochem.* **125,** 131–138.
6. Schlabach, T. D., and Wehr, T. C. (1982) Fluorescent techniques for the selective detection of chromatographically separated peptides. *Anal. Biochem.* **127,** 222–233.
7. Lai, C. Y. (1977) Detection of peptides by fluorescence methods, in *Methods of Enzymology* (eds. Hirs, C. H. W. and Timasheff, S. N.,) vol. 47, pp. 236–243. Academic Press, London.
8. Mahoney, W. C., and Hermodson, M. A. (1980) Separation of large denatured peptides by reverse-phase high performance liquid chromatography. Trifluoroacetic acid as a peptide solvent. *J. Biol. Chem.* **255,** 11,199–11,203.

Chapter 6

SDS Polyacrylamide Gel Electrophoresis of Proteins

B. J. Smith

Institute of Cancer Research, Chester Beatty Laboratories, Royal Cancer Hospital, Fulham Road, London, United Kingdom

Introduction

Probably the most widely used of techniques for analyzing mixtures of proteins is SDS polyacrylamide gel electrophoresis. In this technique, proteins are reacted with the anionic detergent, sodium dodecylsulfate (SDS, or sodium lauryl sulfate) to form negatively charged complexes. The amount of SDS bound by a protein, and so the charge on the complex, is roughly proportional to its size. Commonly, about 1.4 g SDS is bound per 1 g protein, although there are exceptions to this rule. The proteins are generally denatured and solubilized by their binding of SDS, and the complex forms a prolate elipsoid or rod of a length roughly proportionate to the protein's molecular weight. Thus, proteins of either acidic or basic p*I* form negatively charged complexes that can be separated on the bases of differences in charges and sizes by electrophoresis through a sieve-like matrix of polyacrylamide gel.

This is the basis of the SDS gel system, but it owes its popularity to its excellent powers of resolution that derive from the use of a "stacking gel." This system employs the principles of isotachophoresis, which effectively concentrates samples from large volumes (within reason) into very small zones, that then leads to better separation of the different species. The system is set up by making a "stacking gel" on top of the "separating gel," which is of a different pH. The sample is introduced to the system at the stacking gel. With an electric field applied, ions move towards the electrodes, but at the pH prevailing in the stacking gel, the protein–SDS complexes have mobilities intermediate between the Cl^- ions (present throughout the system) and glycinate ions (present in the reservoir buffer). The Cl^- ions have the greatest mobility. The following larger ions concentrate into narrow zones in the stacking gel, but are not effectively separated there. When the moving zones reach the separating gel, their respective mobilities change in the pH prevailing there and the glycinate ion front overtakes the protein–SDS complex zones to leave them in a uniformly buffered electric field to separate from each other according to size and charge. More detailed treatments of the theory of isotachophoresis and electrophoresis generally are available in the literature (e.g., *1*).

The system of buffers used in the gel system described below is that of Laemmli (*2*), and is used in a polyacrylamide gel of slab shape. This form allows simultaneous electrophoresis of more than one sample, and thus is ideal for comparative purposes.

Materials

1. The apparatus required is available commercially or can be made in the workshop, but generally conforms to the design by Studier (*3*). The gel is prepared and run in a narrow chamber formed by two glass plates separated by spacers of perspex or other suitable material,

as shown in Fig. 1. The spacers are longer than the glass plates (so that they are easy to remove), are about 1 cm wide, and are of any thickness (say, 0.5–1 mm for a thin gel). The sample wells into which samples are loaded are formed by a template "comb" that extends across the top of the gel and is of the same thickness as the spacers. Typically, the "teeth" on this comb will be 1 cm long, 2–10 mm wide, and separated by about 3 mm. The chamber is sealed with white petroleum jelly (Vaseline). A dc power supply is also required.

2. Stock solutions. Chemicals should be analytical reagent (Analar) grade and water should be distilled. Stock solutions should all be filtered. Cold solutions should be warmed to room temperature before use.

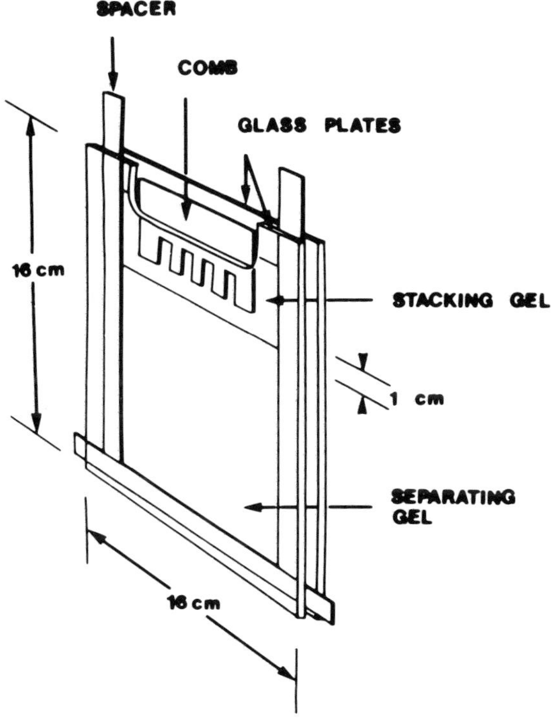

Fig. 1. The construction of a slab gel, showing the positions of the glass plates, the spacers, and the comb.

(i) Stock acrylamide solution (total acrylamide content, %*T* = 30% w/v, ratio of crosslinking agent to acrylamide monomer, %C = 2.7% w/w)

Acrylamide	73 g
Bis acrylamide	2 g

Dissolve the above and make up to 250 mL in water. This stock solution is stable for weeks in brown glass, at 4°C.

(ii) Stock separating gel buffer.

SDS	1.0 g
"Tris" buffer [2-amino-2-(hydroxymethyl)-propane-1,3-diol]	45.5 g

Dissolve the above in less than 250 mL of water, adjust the pH to 8.8 with HCl, and make the volume to 250 mL. This stock solution is stable for months at 4°C.

(iii) Stock ammonium persulfate.

Ammonium persulfate	1.0 g

Dissolve in 10 mL of water. This stock solution is stable for weeks in brown glass at 4°C.

(iv) Stock stacking gel buffer.

SDS	1.0 g
"Tris" buffer	15.1 g

Dissolve the above in less than 250 mL of water, adjust the pH to 6.8 with HCl, and make up to 250 mL. Check the pH before use. This stock solution is stable for months at 4°C.

(v) Reservoir buffer (0.192*M* glycine, 0.025*M* Tris, 0.1% w/v SDS).

Glycine	28.8 g
"Tris" buffer	6.0 g
SDS	2.0 g

Dissolve the above and make to 2 L in water. The solution should be at about pH 8.3 without adjustment. This solution is readily made fresh each time.

(vi) Stock (double strength) sample solvent:

SDS	0.92 g
β-Mercaptoethanol	2 mL
Glycerol	4.0 g
"Tris" buffer	0.3 g
Bromophenol Blue (0.1% w/v solution in water)	2 mL

Dissolve the above in less than 20 mL of water, adjust the pH to 6.8 with HCl, and make to 20 mL. Check the pH before use. Exposed to oxygen in the air, the reducing power of the β-mercaptoethanol wanes with time. Periodically (after a few weeks) add extra agent or renew the solution. This stock solution is stable for weeks at 4°C.

(vii) Protein stain.

Coomassie Brilliant Blue R250	0.25 g
Methanol	125 mL
Glacial acetic acid	25 mL
Water	100 mL

Dissolve the Coomassie dye in the methanol component first, then add the acid and water. If dissolved in a different order, the dye's staining behavior may differ. The stain is best used when freshly made. For best results do not reuse the stain—its efficacy declines with use. If this dye is not available, use the equivalent dye PAGE blue 83.

(viii) Destaining solution.

Methanol	100 mL
Glacial acetic acid	100 mL
Water	800 mL

Mix thoroughly. Use when freshly made.

Method

1. Thoroughly clean and dry the glass plates and three spacers, then assemble them as shown in Fig. 1, with the spacers set 1–2 mm in from the edges of the glass

plates. Hold the construction together with bulldog clips. White petroleum jelly (melted in a boiling water bath) is then applied around the edges of the spacers to hold them in place and seal the chamber. Clamp the chamber in an upright, level position.

2. A sufficient volume of separating gel mixture (say 30 mL for a chamber of about 14 × 14 × 0.1 cm) is prepared as follows. Mix the following:

Stock acrylamide solution	15 mL
Distilled water	7.5 mL

Degas on a water pump, and then add:

Stock separating gel buffer	7.5 ml
Stock ammonium persulfate solution	45 μL
N, N, N', N'-tetramethyl-ethylenediamine (TEMED)	15 μL

Mix gently and use immediately (because polymerization starts when the TEMED is added). The degassing stage removes oxygen, which inhibits polymerization by virtue of mopping up free radicals, and also discourages bubble formation when pouring the gel.

3. Carefully pipet or pour the freshly mixed solution into the chamber without generating air bubbles. Pour to a level about 1 cm below where the bottom of the well-forming comb will come when it is in position. Carefully overlayer the acrylamide solution with butan-2-ol without mixing (to eliminate oxygen and generate a flat top to the gel). Leave the mixture until it is set (0.5–1.5 h).

4. Prepare stacking gel (5 mL) as follows. Mix the following:

Stock acrylamide solution	0.75 mL
Distilled water	3 mL

Degas on a water pump, then add:

Stock stacking gel buffer	1.25 mL
Stock ammonium persulfate solution	15 μL
TEMED	5 μL

Mix gently and use immediately.

Pour off the butan-2-ol from the polymerized separating gel, wash the gel top with water and then a little stacking gel mixture, and fill the gap remaining in the chamber with the stacking gel mixture. Insert the comb and allow the gel to stand until set (about 0.5–1 h).

5. When the stacking gel has polymerized, remove the comb without distorting the shapes of the well. Remove the clips holding the plates together, and install the gel in the apparatus. Fill apparatus with reservoir buffer. The reservoir buffer can be circulated between anode and cathode reservoirs, to equalize their pH values. The buffer can also be cooled (by circulating it through a cooling coil in ice), so that heat evolved during electrophoresis is dissipated and does not affect the size or shape of protein zones (or bands) in the gel. Push out the bottom spacer from the gel and remove bubbles from both the top and underneath of the gel, for they could partially insulate the gel and distort electrophoresis. Check the electrical circuit by turning on the power (dc) briefly, with the cathode at the stacking gel end of the gel (i.e., the top). Use the gel immediately.
6. While the gel is polymerizing (or before making the gel), prepare samples for electrophoresis. A dry sample may be dissolved directly in single-strength sample solvent (i.e., the stock solution diluted twofold with water) or dissolved in water and diluted with one volume of stock double-strength sample solvent. The concentration of sample in the solution should be such as to give a sufficient amount of protein in a volume not greater than the size of the sample well. Some proteins may react adequately with SDS within a few minutes at room temperature, but as a general practice, heat sample solutions in boiling water for 2 min. Cool the sample solution before loading it. The bromophenol blue dye indicates when the sample solution is acidic by turning yellow. If this happens, add a little NaOH, enough to just turn the color blue.
7. Load the gel. Take up the required volume of sample solution in a microsyringe or pipet and carefully inject

it into a sample well through the reservoir buffer. The amount of sample loaded depends upon the method of its detection (see below). Having loaded all samples without delay, start electrophoresis by turning on power (dc). On a gel of about 0.5–1 mm thickness and about 14 cm length, an applied voltage of about 150 V gives a current of about 20 mA or so (falling during electrophoresis if constant voltage is employed). The bromophenol blue dye front takes about 3 h to reach the bottom of the gel. Greater voltage speeds up electrophoresis, but generates more heat in the gel.

8. At the end of electrophoresis (say, when the dye front reaches the bottom of the gel), protein bands in the gel may be visualized by staining. Remove the gel from between the glass plates and immerse it in the protein stain immediately (although delay of an hour or so is not noticeably detrimental in a gel of 15%T). The gel is left there with gentle agitation until the dye has penetrated the gel (about 1.5 h for 15%T gels of 0.5–1 mm thickness). Dye that is not bound to protein is removed by transferring the gel to destaining solution. After about 24 h, with gentle agitation and several changes of destaining agent, the gel background becomes colorless and leaves protein bands colored blue, purple, or red. Coomassie Brilliant Blue R250 and PAGE blue 83 each visibly stain as little as 0.1–1 μg of protein in a band of about 1 cm width.

Notes

1. The reducing agent in the sample solvent reduces intermolecular disulphide bridges and so destroys quarternary structure and separates subunits, and also oxidizes intramolecular disulfide bonds to ensure maximal reaction with SDS. The glycerol is present to increase the density of the sample, to aid the loading of it onto the gel. The bromophenol blue dye also aids loading of the sample, by making it visible, and indicates the position of the front of electrophoresis in the gel. The dye also indicates when the sample solution is acidic by turning yellow.
2. The polymerization of acrylamide and bisacrylamide is

initiated by the addition of TEMED and persulfate. The persulfate activates the TEMED and leaves it with an unpaired electron. This radical reacts with an acrylamide monomer to produce a new radical that reacts with another monomer, and so on to build up a polymer. The bis acrylamide is incorporated into polymer chains this way and so forms crosslinks between them.

3. The gel system described is suitable for electrophoresis of proteins in the M_r range of 10,000–100,000. Smaller proteins move at the front or form diffuse, fast-moving bands, whereas larger proteins hardly enter the gel, if at all. Electrophoresis of larger proteins requires gels of larger pore size, which are made by dilution of the stock acrylamide solution (reduction of %T) or by adjustment of %C (the smallest pore size is at 5%C, whatever the %T). The minimum %T is about 3%, useful for separation of proteins of molecular weights of several millions. Such low %T gels are extremely weak and may require strengthening by the inclusion of agarose to 0.5% w/v. Smaller pore gels, for electrophoresis of small proteins, are prepared by increasing %T and adjustment of %C. Such adjustment of %T and %C may be found empirically to improve resolution of closely migrating species.

   A combination of large and small pore gels, suitable for electrophoresis of mixtures of proteins of wide-ranging sizes, can be made in a gradient gel, prepared with use of a gradient-making apparatus when pouring the separating gel (see Chapter 7).

4. Since proteins (or rather, their complexes with SDS) are resolved largely on the basis of differences in their sizes, electrophoretic mobility in SDS gels may be used to estimate the molecular weight of a protein by comparison with proteins of known size [as described in (*1*)]. However, it should be remembered that some proteins have anomalous SDS-binding properties, and hence anomalous mobilities in SDS gels.

5. If necessary, the gel may be stored for 24 h (preferably in the cold) either as the separating gel only, under a buffer of stock-separating gel buffer diluted fourfold in water, or together with the stacking gel with the comb

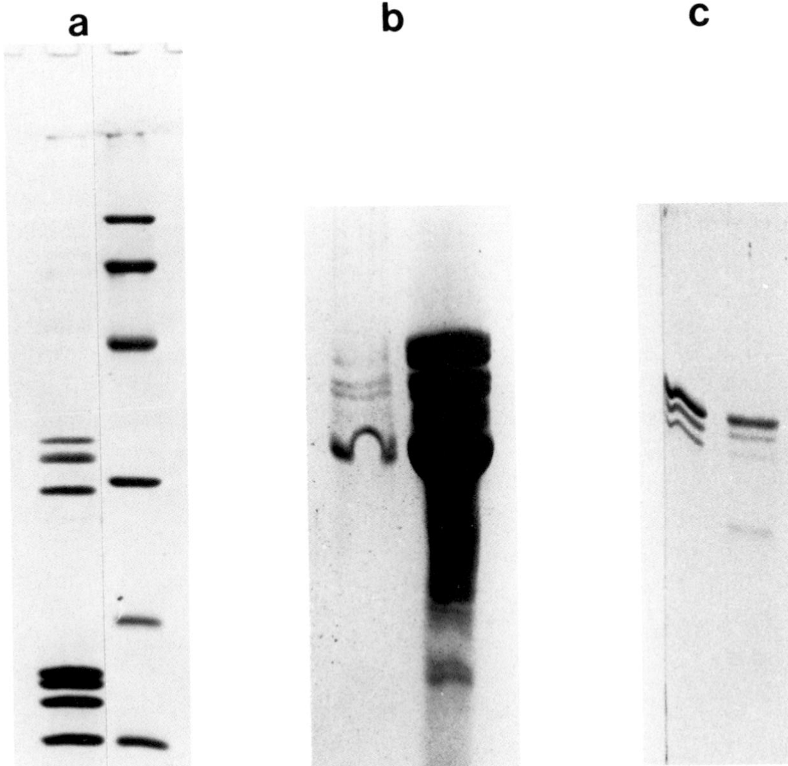

Fig. 2. Examples of proteins electrophoresed on SDS polyacrylamide (15%T) gels and stained with Coomassie Brilliant Blue R 250 as described in the text. Electrophoresis was from top to bottom. (a) Good electrophoresis. Sample, left, 15-μg loading of histone proteins from chicken erythrocyte nuclei. Sample right, a 5-μg loading (total) of molecular weight marker proteins (obtained from Pharmacia). The M_rs are, from top to bottom: phosphorylase b, 94,000; albumin, 67,000; ovalbumin, 43,000; carbonic anhydrase, 30,000; trypsin inhibitor, 20,100; α-lactalbumin, 14,400. (b) Examples of artifacts. Sample, left, the fastest (bottom) band has distorted as it encountered a region of high polyacrylamide density (which arose during very rapid gel polymerization). Sample, right, the effect on protein overloading of increasing a band's size (the sample proteins are as in sample, left). Extreme overloading may also cause narrowing of faster-migrating bands, as has happened here to some extent (cf. fast bands' widths with widths of bands in sample, left). (c) Example of an artifact. Sample, left, the "end well effect" of distortion of the sample loaded into the very end well, not seen in samples in other wells (e.g., sample, right).

Table 1
Some Problems That May Arise During the Preparation and Use of SDS Gels

	Fault	Cause	Remedy
(i)	Failure or a decreased rate of gel polymerization	(a) Oxygen is present. (b) Stock solutions (especially acrylamide and persulfate) are aged	(a) Degas the solutions (b) Renew the stock solutions
(ii)	Formation of a sticky top to the gel	Penetration of the gel by butan-2-ol	Overlayer the gel solution with butan-2-ol without mixing them. Do not leave butan-2-ol to stand on a polymerized gel
(iii)	Poor sample wells		
(a)	The wells are distorted or broken	(a) The stacking gel resists the removal of the comb	(a) Remove the comb carefully or use a gel of lower %T
(b)	The wells contain a loose webbing of polyacrylamide	(b) The comb fits loosely	(b) Replace the comb with a tighter-fitting one
(iv)	Unsatisfactory staining		
(a)	The staining is weak.	(a) The dye is bound inefficiently	(a) Use a more concentrated dye solution, a longer staining time, or a more sensitive stain. The stain solution should contain organic solvent (e.g., methanol), which strips the SDS from the protein to which the dye may then bind

(continued)

Table 1 (*continued*)

	Fault	Cause	Remedy
(b)	The staining is uneven	(b) The dye penetration or destaining is uneven	(b) Agitate the gel during staining and destaining. Increase the staining/destaining time
(c)	Stained bands become decolorized	(c) The dye has been removed from the protein	(c) Restain the gel. Reduce the destaining time or use a dye that stains proteins indelibly, e.g., Procion Navy MXRB (see Chapter 14)
(d)	The gel is marked nonspecifically by the dye	(d) Solid dye is present in the staining solution	(d) Ensure full dissolution of the dye, or filter the solution before using it
(v)	Contaminants are apparent	(a) The apparatus and/or stock solutions are contaminated	(a) Clean or renew them as required
		(b) Nonproteinaceous material in the sample (e.g., nucleic acid) has been stained	(b) Try another stain that will not stain the contaminants.
		(c) Samples have cross-contaminated each other because of their overloading or their sideways seepage between the gel layers	(c) Do not overfill sample wells. Ensure good adherence of the gel layers to each other by thorough washing of the polymerized gel before application of subsequent layers.

(*continued*)

Table 1 (*continued*)

	Fault	Cause	Remedy
(vi)	Protein bands are not sufficiently resolved	(a) Insufficient electrophoresis	(a) Prolong the run
		(b) The separating gel's pore size is incorrect	(b) Alter the %T and/or %C of the separating gel
(vii)	There are small changes in standard proteins' electrophoretic mobilities from time to time	(a) The amounts loaded differ greatly	(a) Keep the loadings roughly similar in size each time
		(b) The constituents of the gel vary in quality from batch to batch or with age	(b) Use one batch of a chemcial for as long as possible. Replace aged stock solutions and reagents
(viii)	Distortion of bands		
(a)	Bands have become smeared or streaked	(a) Proteins in the sample are insoluble or remain aggregated in the sample solvent	(a) Use fresh sample solvent and/or extra SDS and reducing agent in it (especially for concentrated sample solutions)
		There is insoluble matter or a bubble in the gel that has interfered with protein band migration	Filter the stock solutions before use and remove any bubbles from the gel mixtures
		The pore size of the gel is inconsistent	Ensure that the gel solutions are well mixed and that polymerization is not very

(*continued*)

Table 1 (*continued*)

	Fault	Cause	Remedy
			rapid (to slow it down, reduce the amount of persulfate added)
(b)	Protein migration has been uneven (bands are bent)	Part of the gel has been insulated.	(b) Remove any bubbles adhering to the gel before electrophoresis.
		Electrical leakage	Ensure that the side spacers are in place
		Cooling of the gel is uneven (allowing one part of the gel to run more quickly than another)	Improve the cooling of the gel, or reduce the heating by reducing the voltage or ionic strength of buffers
		The band and/or its neighbors are overloaded	Repeat the electrophoresis, but with smaller loadings. Leave gaps (i.e., unloaded sample wells) between neighboring heavily loaded samples. If necessary, alter the system [e.g. see (*vi*)] and so the relative mobilities of bands, so that they do not interfere with each other
		The sample well used was at the very end of the row of	Avoid using the end wells

(*continued*)

Table 1 (*continued*)

Fault	Cause	Remedy
	wells (the "end well effect")	
(c) Bands are not of uniform thickness	(c) The sample was loaded unevenly	(c) Check that the sample well bottoms are straight and horizontal [see (*iii*)]

left in place to prevent drying out.

6. Proteins dissolved in sample solvent are stable for many weeks if kept frozen (at −10°C or below), although repeated freezing and thawing causes protein degradation.
7. The result of electrophoresis in SDS gels ideally has protein(s) as thin, straight band(s) that are well-resolved from other bands. This may not always be so, however. Some faults and their remedies are given in Table 1. Some examples are shown in Fig. 2.
8. Be wary of the dangers of electric shock and of fire, and of the neurotoxic acrylamide monomer.

References

1. Deyl, Z. (1979) *Electrophoresis. A survey of techniques and applications. Part A: Techniques.* Elesevier, Amsterdam.
2. Laemmli, U.K. (1970) Cleavage of structural proteins during the assembly of the head of bacteriophage T4. *Nature* **227,** 680–685.
3. Studier, F. W. (1973) Analysis of bacteriophage T7 early RNAs and proteins on slab gels. *J. Mol. Biol.* **79,** 237–248.

Chapter 7

Gradient SDS Polyacrylamide Gel Electrophoresis

John M. Walker

School of Biological and Environmental Sciences, The Hatfield Polytechnic, Hatfield, Hertfordshire, England

Introduction

The preparation of fixed-concentration polyacrylamide gels has been described in Chapter 6. However, the use of polyacrylamide gels that have a gradient of increasing acrylamide concentration (and hence decreasing pore size) can sometimes have advantages over fixed-concentration acrylamide gels. During electrophoresis in gradient gels, proteins migrate until the decreasing pore size impedes further progress. Once the "pore limit" is reached, the protein banding pattern does not change appreciably with time, although migration does not cease completely. There are three main advantages of gradient gels over linear gels:

1. The advancing edge of the migrating protein zone is retarded more than the trailing edge, thus resulting in a sharpening of the protein bands.
2. The gradient in pore size increases the range of molecular weights that can be fractionated in a single gel run.
3. Proteins with close molecular weight values are more likely to separate in a gradient gel than a linear gel.

The usual limits of gradient gels are 3–30% acrylamide in linear or concave gradients. The choice of range will of course depend on the size of proteins being fractionated. The system described here is for a 5–20% linear gradient using SDS polyacrylamide gel electrophoresis. The theory of SDS polyacrylamide gel electrophoresis has been described in Chapter 6.

Materials

1. Stock acrylamide solution (30% acrylamide, 0.8% bisacrylamide). Dissolve 75 g of acrylamide and 2.0 g of *N,N'*-methylene bisacrylamide in about 150 ml of water. Filter and make the volume to 250 ml. Store at 4°C. The solution is stable for months.
2. Buffers: (a) 1.875*M* Tris-HCl, pH 8.8
(b) 0.6*M* Tris-HCl, pH 6.8
} Store at 4°C
3. Ammonium persulfate solution (5%, w/v). Make fresh as required.
4. SDS solution (10% w/v). Stable at room temperature. In cold conditions, the SDS can come out of solution, but may be redissolved by warming.
5. *N,N,N',N'*-Tetramethylene diamine (TEMED).
6. Gradient forming apparatus (*see* Fig. 1). Reservoirs with dimensions of 2.5 cm id and 5.0 cm height are suitable. The two reservoirs of the gradient former should be linked by flexible tubing to allow them to be moved independently. This is necessary since although equal volumes are placed in each reservoir, the solutions differ in their densities and the relative positions of A and B have to be adjusted to balance the two solutions when the connecting clamp is opened.

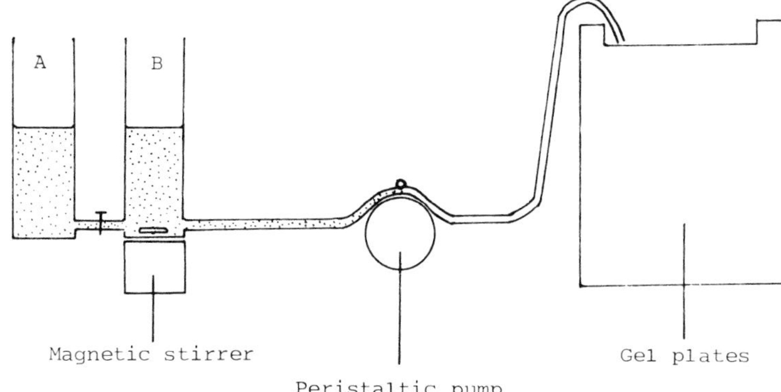

Fig. 1. Diagram of an apparatus for forming gradient gels.

Method

1. Prepare the following solutions, A and B.

	Solution A, mL	Solution B, mL
Tris, pH 8.8	3.0	3.0
Water	9.3	0.6
Stock acrylamide, 30%	2.5	10.0
10% SDS	0.15	0.15
Ammonium persulfate (5%)	0.05	0.05
Sucrose	—	2.2 g (equivalent to 1.2 mL volume)

2. Degas each solution under vacuum for about 30 s and then, when you are ready to form the gradient, add TEMED (10 μL) to each solution.
3. Once the TEMED is added and mixed in, pour solutions A and B into the appropriate reservoirs (see Fig. 1).
4. *With the stirrer stirring,* fractionally open the connection between A and B and adjust the relative heights of A and B such that there is no flow of liquid between the two reservoirs (easily seen because of the difference in densities). Do not worry if there is some mixing between reservoirs—this is inevitable.

5. When the levels are balanced, completely open the connection between A and B, turn the pump on, and fill the gel apparatus by running the gel solution down one edge of the gel slab. Surprizingly, very little mixing within the gradient occurs using this method. A pump speed of about 5 mL/min is suitable. If a pump is not available, the gradient may be run into the gel under gravity.
6. When the level of the gel reaches about 3 cm from the top of the gel slab, connect the pump to distilled water and overlay the gel with 3–4 mm of water.
7. The gradient gel is now left to set for 30 min. Remember to rinse out the gradient former before the gel sets in it.
8. Prepare a stacking gel by mixing the following:

Tris pH 6.8	1.0 mL
Stock acrylamide	1.35 mL
Water	7.5 mL
10% SDS	0.1 mL
Ammonium persulfate (5%)	0.05 mL

9. Degas this mixture under vacuum for 30 s and then add TEMED (10 μL).
10. Pour off the water overlayering the gel and wash the gel surface with about 2 mL of stacking gel solution and then discard this solution.
11. The gel slab is now filled to the top of the plates with stacking gel solution and the well-forming comb placed in position (*see* Chapter 6).
12. When the stacking gel has set (~ 15 min), carefully remove the comb. The gel is now ready for running. The conditions of running and sample preparation are exactly as described for SDS gel electrophoresis in Chapter 6.

Notes

1. The total volume of liquid in reservoirs A and B should be chosen such that it approximates to the volume available between the gel plates. However, allowance

must be made for some liquid remaining in the reservoirs and tubing.

2. As well as a gradient in acrylamide concentration, a density gradient of sucrose (glycerol could also be used) is included to minimize mixing by convectional disturbances caused by heat evolved during polymerization. Some workers avoid this problem by also including a gradient of ammonium persulfate to ensure that polymerication occurs first at the top of the gel, progressing to the bottom. However, we have not found this to be necessary in our laboratory.

Chapter 8

Acetic Acid–Urea Polyacrylamide Gel Electrophoresis of Proteins

B. J. Smith

Institute of Cancer Research, Royal Cancer Hospital, Chester Beatty Laboratories, Fulham Road, London, England

Introduction

In SDS polyacrylamide gel electrophoresis, proteins are separated essentially on the basis of their sizes, by the sieving effect of the polyacrylamide gel matrix (*see* Chapter 6). In the absence of SDS, the proteins would still be subject to the sieving effect of the gel matrix, but their charges would vary according to their amino acid content. This is because the charge on a protein at any particular pH is the sum of the charges prevailing on the side chain groups of its constituent amino acid residues, and the free amino and carboxyl groups at its termini (although these are relatively trivial in anything other than a very small peptide).

Thus, in an ionic detergent-free gel electrophoretic system, both the molecular size and charge act as bases for effective protein separation. The pH prevailing in such a system might be anything, but is commonly about pH 3. Since the pK_a values of the side chain carboxyl groups of aspartic and glutamic acids are about 3.8 and 4.2, respectively, even these amino acids will contribute little to the negative charge on a protein at this pH. Thus at pH 3, all proteins are likely to be positively charged and to travel towards the cathode in an electric field.

In such an acid–polyacrylamide gel electrophoresis system, two proteins of similar size but different charge may be separated from each other. Since SDS gels may be unable to achieve this end, these two electrophoresis systems usefully complement each other for analysis of small amounts of proteins. Proteins that might be usefully studied in the acid–gel system are minor primary structure variants (of slightly different charge), or modified forms of the same protein. Thus, a protein that has had some threonine or serine side chains phosphorylated, or lysine side chains acetylated, will be more acidic (or less basic) than the unmodified form of the same protein, and so will have a different electrophoretic mobility in the appropriate acid–gel system (for instance, see the acetylated derivatives of H4 in Fig. 1).

Commonly, the hydrogen bond-breaking agent urea is added to the simple acid–gel electrophoresis system in amounts traversing its entire range of solubility. This denaturant increases the frictional coefficient of proteins and so alters their electrophoretic mobilities. This has often proved useful in obtaining optimal resolution of proteins of interest and so urea is included in the system described below, which uses 2.5*M* urea. The system is buffered to about pH 3 with acetic acid, and is similar to the system described by Panyim and Chalkley (*1*).

Materials

1. The apparatus required for running slab gels is available commercially or may be made in the workshop, but is usually of the type described by Studier (*2*). The gel is cast and used in a chamber formed between two glass plates, as are SDS gels (for further details, *see* Chapter 6). A dc power supply is required.

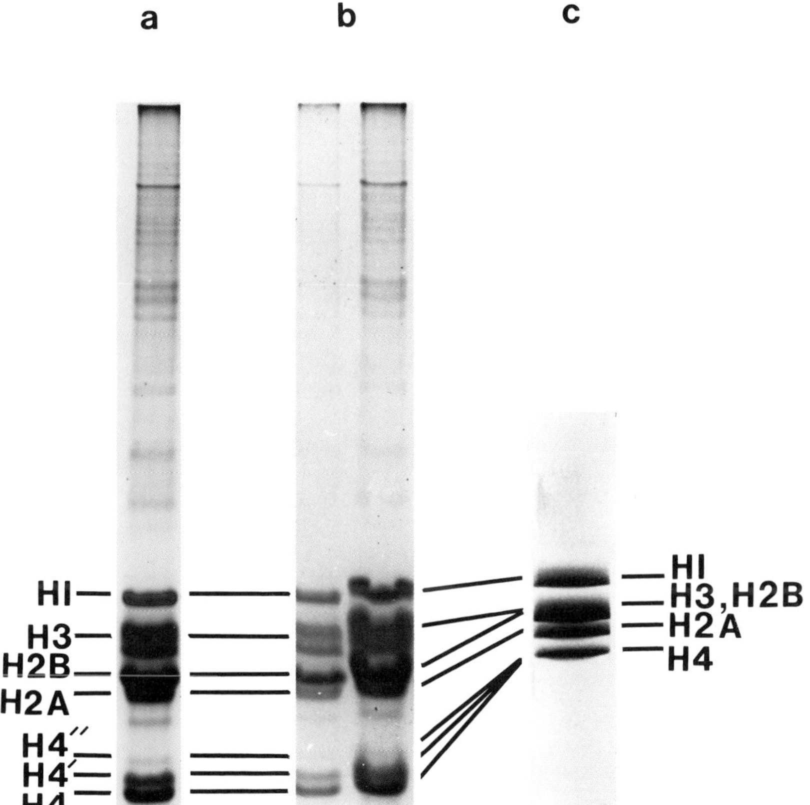

Fig. 1. Examples of electrophoresis on acetic acid (0.9*M*, pH 3)–polyacrylamide (20%T, 1.5%C) gels. (a) Slab gel containing 2.5*M* urea, stained with Coomassie Brilliant Blue R250, as described in the text. Sample: 8 μg of a mouse liver nuclei extract. The histones are identified. The H3 band probably also contains another protein, H1°. Note the mono- and the faint band of diacetylated forms of H4 (H4′ and H4″, resp.) migrating behind the non-acetylated H4. (b) Slab gel containing 2.5*M* urea, stained with Procion Navy MXRB as described in Chapter 14. Sample: mouse liver nuclei extract. Left, loading 8 μg; right, loading 24 μg. Note the different sensitivity of the stain [cf. (a)] and the distortion of bands that occurs with heavier loading. (c) Rod gel (5 mm diameter) containing *no* urea, stained with Procion Navy MXRB as in (*10*). Sample: 50 μg of pig thymus histones. Note alteration of H2B mobility relative to other histones because of the omission of urea.

2. Stock solutions. Use analytical grade (Analar) reagents and distilled water. Filter stock solutions and warm to room temperature before use.

   (i) *Stock acrylamide solution* (total acrylamide content, %T = 30% w/v, ratio of crosslinking agent to acrylamide monomer, %C = 1.5% w/w).

Acrylamide	73.8 g
Bis acrylamide	1.1 g

   Dissolve the above and make to 250 mL in water. Filter the solution before use. This stock solution is stable for weeks in brown glass at 4°C.

   (ii) *Stock ammonium persulfate.*

Ammonium persulfate	1 g

   Dissolve in 10mL of water. This stock solution is stable for weeks in brown glass at 4°C.

   (iii) *Reservoir buffer,* pH 3 (0.9*M* acetic acid).

Acetic acid (glacial)	51.5 mL

   Make up to 1L with water. Can be stored, but is readily made fresh each time.

   (iv) *Sample solvent*

HCl (1*M*)	1 mL
β-mercaptoethanol	0.5 mL
Urea	5.4 g
Pyronin Y (0.4% w/v solution in water)	0.5 mL

   Add 4.5 mL distilled water to the above and fully dissolve the urea. The final volume is 10 mL. Although it is probably best to use this solution when it is fresh, it may be stored frozen at −20°C for weeks without apparent adverse effect.

   (v) *Protein stain*

Coomassie Brilliant Blue R250	0.25 g
Methanol	125 mL
Acetic acid (glacial)	25 mL
Water	100 mL

Dissolve the dye in the methanol, then add the acid and water. If dissolved up in a different order the dye's staining behavior may differ. The stain is best used when freshly made. For best results do not reuse the stain—its efficacy declines with use. If this dye is not available use the equivalent dye PAGE blue 83.

(vi) *Destaining solution*:

Methanol	100 mL
Acetic acid (glacial)	100 mL
Water	800 mL

Mix thoroughly. Make when required and use fresh.

Method

1. Assemble clean glass plates and spacers into the form of a chamber (as described for SDS gels, Chapter 6), and seal it with molten white petroleum jelly. Clamp it in an upright, level position.
2. Prepare a sufficient volume of separating gel mixture (30 mL for a chamber of about 14×14×0.1 cm allows for some wastage), as follows. Mix the following:

Stock acrylamide solution	20 mL
Urea	4.5 g
Glacial acetic acid	1.57 mL
N,N,N′,N′-tetramethyl-ethylenediamine (TEMED)	100 μL

Make to 29.45 mL with distilled water, degas on a water pump, then add stock ammonium persulfate solution (0.55 mL).

Mix gently and use immediately because polymerization starts when the persulfate is added. The degassing removes oxygen, which can inhibit polymerization (by virtue of mopping up free radicals), and also discourages bubble formation when pouring the gel.

3. Carefully pipet or pour freshly made solution into the chamber and remove any bubbles present. Pour to a level about 0.5 cm below where the bottom of the well-forming template ("comb") will come when it is in position. Carefully overlayer the acrylamide solution with butan-2-ol, without mixing. This insulates the solution

from oxygen and generates a flat top to the gel. The acrylamide should polymerize in an hour or so at room temperature.

4. Prepare the upper gel layer (5 mL) as follows. Mix the following:

Stock acrylamide solution	1.25 mL
Urea	0.75 g
Glacial acetic acid	78.2 μL
TEMED	25 μL

   Make to 4.75 mL with water, degas on water pump, then add stock ammonium persulfate solution (0.24 mL). Mix gently and use immediately. Pour off the butan-2-ol from the polymerized separating gel, wash the gel top with water and then a little of the upper gel mixture, and then fill the gap remaining in the chamber with upper gel mixture. Insert the well-forming comb and allow to stand until set (about 0.5–1 h).

5. When the upper gel has set, remove the comb without breaking or distorting the sample wells. Install the gel in the apparatus and fill the reservoirs with buffer (0.9*M* acetic acid). The gel may be run at room temperature without buffer circulation or cooling. Cooling tends to slow up the rate of electrophoresis and so the advantage of the decreased rate of protein diffusion (which causes band widening) is counteracted by increased diffusion occurring during the longer time required to complete the run. Push out the bottom spacer from the chamber and, with the cathode at the bottom end of the gel, turn on the dc power supply, to give about 180 V. Continue this electrophoresis without added samples ("pre-electrophoresis") at a constant voltage until the current falls to a steady level (say, from 25 to 5 mA for a 0.5–1 mm-thick slab of this sort), or alternatively, at constant current until the voltage increases to a steady level. This process may take about 5 h, but may be done conveniently overnight. The pre-electrophoresed gel may be stored under fresh 0.9*M* acetic acid for at least several days at room temperature.

6. While the gel is polymerizing, prepare samples for electrophoresis. Dry samples will be dissolved directly in sample solvent or aqueous solutions may be diluted

with not less than one volume of sample solvent. The concentration of protein in the solution should be as great as possible so that the volume of solution loaded onto the gel is as small as possible.

7. Prepare to load the gel. Use fresh reservoir buffer in the apparatus. Take up the required volume of sample solution in a microsyringe or pipet and carefully inject it into a sample well, loading it through the reservoir buffer without mixing. The amount of sample loaded depends partly on the sensitivity of the method of detection (see below), but in any case it is often found that good, even, and straight bands are obtained most frequently when samples are lightly loaded, say about 1 μg/1 cm wide band in a 0.5–1 mm thick gel. Having loaded the samples without delay, start electrophoresis at 180 V. The pyronin Y will take 4–5 h to migrate to the bottom of a thin, 14 cm-long gel of the type described here. Decreasing the voltage will prolong the run, whereas an increased voltage will generate more heat, which may not benefit the appearance of the protein bands.
8. At the end of electrophoresis, remove the gel and immerse it in protein stain for at least several hours with gentle agitation. The dye that is then not bound to the protein bands may be removed by washing the gel in several changes of destaining solution. After destaining, protein bands are seen to be colored blue to purplish–red. Excessive destaining will decolorize these protein bands, but they may be restained. Coomassie Brilliant Blue R250 and PAGE Blue 83 each visibly stain as little as 0.1–1 μg of protein in a 1 cm-wide band.

Notes

1. The purpose of the upper, weak polyacrylamide, gel is to provide a medium in which sample wells can be formed and from which the well-forming comb can be readily removed (20%T gel tends to break when the comb is removed). Since the upper gel contains weaker acid than does the separating gel, its pH is slightly higher. However, the purpose of this design

is that the upper gel has lower conductivity than the rest of the system and when the electric field is applied this has a small band-sharpening effect similar to that produced by the "stacking gel" used in SDS gels (*see* Chapter 6).

2. A more typical stacking gel for 0.9*M* acetic acid/urea gels has been described by Spiker (*3*). It has been found in this laboratory that although bands may have an improved appearance, they may instead become smeared.
3. Addition of TEMED and persulfate to the gel mixture initiates its polymerization. This occurs by their interaction and formation of a TEMED radical that reacts with an acrylamide monomer. This in turn produces a radical that reacts with another acrylamide monomer, or occasionally one half of a bis acrylamide molecule. Incorporation of bis acrylamide molecules into different chains forms crosslinks between them.
4. The 9*M* urea in the sample solvent has two functions. First, to disrupt aggregates, and second, to increase the density of the solution (which aids in the loading of the sample beneath the less dense reservoir buffer). The β-mercaptoethanol that is also present reduces inter- and intramolecular disulfide bonds and so, together with the urea, destroys higher-order protein structures. The pyronin Y dye is present to aid sample application by making it visible, and also to mark the approximate position of the front of electrophoresis, near which it runs.
5. The "pre-electrophoresis" treatment of the gel before addition of the samples removes persulfate and other ions that would otherwise slow up the rate of sample electrophoresis and also spoil the resolution of protein bands.
6. As mentioned above, protein bands in this system are often slightly misshapen, and this problem is exacerbated by heavy loading (*see* Fig. 1). Thus, slab gels may give unsatisfactory resolution for quantification purposes (*see* Chapter 14). As an alternative, gels may be made in the forms of rods by casting them in tubes of glass. The result has the proteins as discs running through the rod (e.g., *see* Fig. 1). These discs suffer less from distortion than do bands on slabs. The

separating gel for rod gels is prepared as for slab gels, and polymerized in glass tubes that have been siliconized before-hand. No upper gel is added, but samples are applied directly onto the separating gel. The gel may be extruded from the glass tube, as described in (*4*), or removed after cracking the glass in a vice or with a hammer. The apparatus required for running rod gels is available commercially or may be made to a simple design (*see* ref. *4*).

7. The 20%T gel described is suitable for electrophoresis of smaller proteins (say, of M_r below 50,000). For larger proteins a gel of larger pore size is more convenient by virtue of allowing greater mobility. The pore size may be increased by reduction of %T by simple dilution of the stock acrylamide solution, or by adjustment of %C (which gives the smallest pores when at 5%). It may be found that adjustment of the gel's pore size in this way may also improve the resolution of proteins of interest.
8. If the resolution of the proteins of interest is not good enough, then alteration of pH of the system may be beneficial, by virtue of altering their respective charges by titration of the side chain groups (e.g., *see* ref. *1*). The literature describes polyacrylamide gel systems of various pH values, e.g., pH 4.5 (*5,6*); pH 7.1 (*7*); pH 8.9 (*8*).
9. The concentration of urea in the gel may be altered, to alter the relative mobilities of proteins in the system. The effects of altering the urea concentration have to be determined empirically. An example is shown in Fig. 1, in which it may be seen that in the absence of urea (as in 1*M* urea) histone proteins H2B and H3 comigrate, whereas in 2.5*M* urea they are resolved from each other (*see also* ref. *1*).
10. In addition to urea, the non-ionic detergent Triton X 100 (or Triton DF-16) may be added to the gel (*9*). This agent binds to proteins in proportion to their hydrophobicity, and alters their electrophoretic mobilities accordingly. This technique has proved useful in the study of proteins that differ slightly in their hydrophobic character and that are not separated by ordinary acid/urea or SDS gel electrophoresis.

11. The acid–urea system described may be adapted for separation of nondenatured proteins, which can be detected in the gel by their enzymatic properties and prepared (undenatured) from the gel. For this purpose, denaturants (urea and Triton) are omitted and instead of the sample solvent given above, with its urea and reducing agent, the 0.9*M* acetic acid, 30% w/v sucrose sample solvent of Panyim and Chalkley (*1*) may be used.
12. Coomassie Brilliant Blue R250 may not fully penetrate to and stain the center of bands of concentrated protein, especially in thicker slabs and rod gels. Procion Navy MXRB does not suffer this defect, although it is several-fold less sensitive (*see* Fig. 1). The method of staining with Procion Navy described by Goodwin et al. (*10*) is suitable for rod gels, but the heated destaining process is inconvenient for slabs. However, the method described in Chapter 14 gives an equivalent result and is suitable for all forms of gel.
13. Various problems may arise to spoil the otherwise perfect gel. Failure of the gel to polymerize, distortion of sample wells and other problems that are general to the process of polyacrylamide gel electrophoresis are dealt with in Chapter 6. The main failing with the present system in particular is the common, uneven shape of the protein bands especially in conditions of heavy loading (e.g., *see* Fig. 1). As mentioned above, use of a stacking gel may alleviate this problem, but generally speaking light loading (about 1 μg/band or less) and small sample volumes go a long way to promote good results. Avoid using the sample wells at the very ends of the row of wells, for samples in these often suffer distortion during the electrophoresis.
14. Be wary of the dangers of electric shock and of fire, and of the neurotoxic acrylamide monomer.

References

1. Panyim, S., and Chalkley, R. (1969) High resolution acrylamide gel electrophoresis of histones. *Arch. Biochem. Biophys.* **130,** 337–346.

2. Studier, F. W. (1973) Analysis of bacteriophage T7 early RNAs and protein on slab gels. *J. Mol. Biol.* **79,** 237–248.
3. Spiker, S. (1980) A modification of the acetic acid–urea system for use in microslab polyacrylamide gel electrophoresis. *Anal. Biochem.* **108,** 263–265.
4. Gordon, A. H. (1969) Electrophoresis of proteins in polyacrylamide and starch gels, in *Laboratory Techniques in Biochemistry and Molecular Biology,* Vol. 1 (eds. Work, T. S., and Work, E.), pp. 1–149. North Holland, Amsterdam, London.
5. Kistler, W. S., Geroch, M. E., and Williams-Ashman, H. G. (1973) Specific basic proteins from mammalian testes. Isolation and properties of small basic proteins from rat testes and epididymal spermatozoa. *J. Biol. Chem.* **248,** 4532–4543.
6. Kistler, W. S., and Geroch, M. E. (1975) An unusual pattern of lysine rich histone components is associated with spermatogenesis in rat testis. *Biochem. Biophys. Res. Commun.* **63,** 378–384.
7. Hardison, R., and Chalkley, R. (1978) Polyacrylamide gel electrophoretic fractionation of histones. In *Methods in Cell Biology,* Vol. 17 (eds. Stein, G., Stein, J., and Kleinsmith, L. J). Academic Press, New York.
8 Hrkal, Z. (1979) Gel-type techniques. In *Electrophoresis. A survey of techniques and applications* (ed. Deyl, Z.). Elsevier, Amsterdam.
9. Alfageme, C. R., Zweidler, A., Mahowald, A., and Cohen, L. H. (1974) Histones of Drosophila embryos. Electrophoretic isolation and structural studies. *J. Biol. Chem.* **249,** 3729–3736.
10. Goodwin, G. H., Nicolas, R. H., and Johns, E. W. (1977) A quantitative analysis of histone H1 in rabbit thymus nuclei. *Biochem. J.* **167,** 485–488.

Chapter 9

The In Vivo Isotopic Labeling of Proteins for Polyacrylamide Gel Electrophoresis

Jeffrey W. Pollard

MRC Human Genetic Disease Research Group, Department of Biochemistry, Queen Elizabeth College, University of London, Campden Hill, London, England

Introduction

Autoradiography (*see* Chapter 17 and Vol. 2) offers a convenient, quick, and cheap means of quantifying proteins separated by gel electrophoresis.

For metabolic experiments proteins must be labeled with a radioactive isotope in vivo prior to isolation and subsequent electrophoretic analysis. The isotope chosen, of course, must correspond to the question to be investigated, but they are generally ^{3}H, ^{14}C, ^{32}P, and ^{35}S, which are all beta-emitters. The energies and half lives are:

0.018 MeV (12.43 yr), 0.159 MeV (5600 yr), 1.706 MeV (14.3 d), 0.167 MeV (87.4 d), respectively. ^{3}H is a popular isotope for biological use, potentially available at high specific activity, but its low energy precludes its autoradiography directly and it needs to be either treated for fluorography or quantified by gel slicing and scintillation counting. Thus the higher energy ^{14}C- and ^{35}S-labeled amino acids are commonly used as protein labels since they can be directly detected by autoradiography. However, the low specific activity of ^{14}C labeled amino acids often prevents their use, with the result that ^{35}S-methionine or ^{35}S-cysteine labeling, either alone or in combination, are more commonly used. Nevertheless, it is worth noting that, in a recent paper, Bravo and Celis (*1*) were able to detect 28% more proteins by labeling with a mixture of 16 ^{14}C-amino acids than could be detected with ^{35}S-methionine and that such labeling also allows the long term storage of data on gels. Unfortunately, the cost in isotopes is often prohibitive. In this paper methods for labeling both cells in culture and tissues in vitro with ^{35}S-methionine, ^{3}H or ^{14}C-amino acids, and ^{32}P for phosphoproteins are described, with particular emphasis on sample preparation for two-dimensional polyacrylamide gel electrophoresis (Chapter 10). But these techniques may easily be modified to encompass other uses, such as the isotopes ^{59}Fe, ^{125}I, or ^{131}I for specific studies of particular proteins.

Materials

1. Tissue culture medium: I use a medium rich in amino acids, alpha—minimal essential medium, but lacking in methionine and stored at 4°C. Any defined tissue culture medium may be used.
2. Dialyzed fetal calf serum (DFCS): serum is serially dialyzed against two changes of phosphate-buffered saline to remove amino acids and stored at −20°C. Do not store this at 4°C, since endogenous proteolytic activity will result in a relatively high concentration of amino acids.
3. Ca^{2+} and Mg^{2+}-free phosphate-buffered saline (PBS) (0.14*M* NaC1, 2.7 m*M* KC1, 1.5 m*M* KH_2PO_4, 8.1 m*M* Na_2HPO_4).

4. 0.1% (w/v) trypsin in PBS citrate (PBS containing 20 m*M* sodium citrate).
5. Lysis buffer: 9.5*M* urea, 2% v/v Nonidet, 2% v/v ampholines (1.6%, pH range 5–7; 0.4%, pH range 3.5–10), 5% v/v beta-mercaptoethanol. Stored frozen at −20°C in 0.5 mL aliquots (*see* Chapter 10).
6. Isotopes: ^{35}S-Methionine, ^{32}P-orthophosphate, and ^{3}H-amino acids are used as supplied. ^{14}C-amino acids are lyophilized and resuspended in a medium lacking amino acids at 500 μCi/mL.

Method

Cells in Monolayers

1. Cells are plated directly into microwells at about 2000 cells/well in 0.25 mL of growth medium. They are left to attach for at least 5 h, but preferably overnight.
2. The media is removed and the cells washed with 0.5 mL medium lacking methionine and replaced with 0.1 mL medium supplemented with 10% v/v dialyzed fetal calf serum, lacking methionine, but containing 100 μCi ^{35}S-methionine and 1 mg/L unlabeled methionine.
3. The cells are labeled for 20 h at 37°C (wrap the microtiter plates in cling film to prevent evaporation).
4. Following incubation, the medium is removed, the cells washed twice with PBS to remove serum proteins and, if a total cell extract for two-dimensional gel electrophoresis is required, the cells are lysed with 20 μL lysis buffer. The samples may be stored in small vials or in the microtiter wells at −70°C, or loaded directly onto isoelectric focusing gels (*see* Chapter 10).
5. Alternatively, cells may be trypsinized and processed according to the analytical technique required.
6. Cells may also be adequately labeled for up to about 4 h in a medium completely lacking unlabeled methionine. But under these circumstances, equilibrium labeling may not be achieved.
7. It is difficult to maintain tight physiological control of cells growing in microwells and thus in some experiments where better control of cell physiology is required, cells growing in 25 mL flasks may be labeled in

2 mL of medium containing 10% DFCS and methionine at 1 mg/L containing 100–200 μCi of ^{35}S-methionine. Bottles are occasionally agitated to prevent desiccation of the cells. After labeling, the cells are washed twice with PBS and may be collected either by scraping with a rubber policeman, or trypsinized and processed as before.

Cells Growing in Suspension

1. Cells growing in suspension are collected by centrifugation, washed in methionine-free medium, and regained by centrifugation.
2. The cells are resuspended at about 25% of their saturation density in 15 mL Falcon plastic snap-cap tubes previously gassed with a 5% CO_2–95% air mixture in medium containing 10% DFCS lacking methionine, but supplemented with 400 μCi/mL ^{35}S-methionine and agitated with a small magnetic flea.
3. Cells are labeled for 1–4 h at 37°C, collected by centrifugation, washed twice with PBS, and processed as before.

Tissues

1. The tissue to be labeled is excised, blotted briefly onto filter paper, and chopped into small pieces (approximately 2 mm diameter).
2. These pieces are placed into glass scintillation vials, gassed with a 5% CO_2–95% air mixture with 1 mL of tissue culture medium containing 10% dialyzed fetal calf serum, but lacking methionine supplemented with 200 μCi/mL ^{35}S-methionine.
3. The vials are placed in a shaking water bath at 37°C and labeling is performed for 1 h; thereafter, the sample is processed as appropriate.

Notes

1. Accurate protein synthetic rates over short periods may be determined concurrently in parallel flasks by measuring the rate of incorporation of a mixture of

three ^{14}C-labeled essential amino acids into acid insoluble counts per cell.

2. The handling of cells to be labeled with ^{3}H or ^{14}C-amino acids or ^{32}P-orthophosphate, is identical. The ^{3}H or ^{14}C-amino acids are exposed to cells in 0.1 mL of medium at 500 µCi/mL in medium lacking the appropriate amino acids or to ^{32}P-orthophosphate at 2 mCi/mL in medium lacking phosphate.
3. To determine the number of counts incorporated, a small aliquot (1–5 µL) should be removed from the final sample, added to 0.5 mL of water containing 10 µg of bovine serum albumin, and precipitated with 0.5 mL of ice-cold 20% (w/v) TCA. The precipitate is allowed to develop for 10 min and then collected onto Whatman glass fiber discs with four washings of 5 mL of 5% (w/v) TCA. The discs are finally washed with ethanol, dried, and either digested overnight with a tissue solubilizer for ^{3}H samples, and counted in a compatible scintillant, or counted directly in a scintillation counter. ^{32}P may alternatively be determined by Cerenkov counting.
4. It must be remembered that mammalian cells show large changes in the synthesis of proteins when growth conditions are changed. Thus, even amino acid deprivation results in a reduction of the initiation rate of ribosomes onto mRNA, with the result of preferential synthesis of proteins whose mRNAs have high intrinsic rates of initiation. It is also worth repeating that accurate protein synthesis measurements may be achieved only when very small amounts of radioactive precursors are added to reduced volumes of the same conditioned growth medium containing large amounts of unlabeled precursors that has been removed from the growing culture (*see* ref. *2* for a full discussion). These above conditions are clearly not met when fresh medium, often lacking methionine, is used to label proteins for gel electrophoresis. Considerable care should therefore be exercised in standardizing growth and labeling conditions for any experiments involving comparison of different samples. Often a compromise has to be effected between high levels of incorporation and the physiological constraints of maintaining precursor pool sizes.

5. The only other problems that may be encountered, providing care is taken over sterility, pH, and temperature regulation, is the toxicity of isotopes. This is rarely a problem for a day's labeling, but could potentially be so if proteins are labeled for longer. Fibroblasts will survive in 0.5 mCi ^{32}P/mL in medium containing 0.2 m*M* phosphate and in mutant isolation, 1.5 dpm of ^{3}H-amino acid incorporation per cell is considered lethal, but only after a period in the cold for accumulation of radioactive damage. Thus, toxicity will not be a problem using the above labeling procedures.
6. The procedures given above will result in ^{35}S-methionine-labeled proteins with specific activities of around 10^5 cpm/μg of protein. But obviously the optimal conditions and resultant specific activity of proteins will depend on the cell type and growth conditions. Nevertheless, the methods are applicable to a wide range of cells in culture and tissues. It is also worth noting that, following slab gel electrophoresis, a 1 mm^2 spot containing 3 dpm of ^{35}S-methionine may be readily detected ($>$0.01 OD above background) after 1 wk.

Acknowledgments

This article was prepared while the author's work was supported by the MRC (UK), CRC and Central Research Fund of the University of London.

References

1. Bravo, R., and Celis, J. E. (1982) Up-dated catalogue of HeLa cell proteins. Percentages and characteristics of the major cell polypeptides labelled with a mixture of 16 ^{14}C-labelled amino acids. *Clin. Chem.* **28,** 766–781.
2. Stanners, C. P., Adams, M. E., Harkins, J. L., and Pollard, J. W. (1979) Transformed cells have lost control of ribosome number through the growth cycle. *J. Cell. Physiol.* **100,** 127–138.

Chapter 10

Two-Dimensional Polyacrylamide Gel Electrophoresis of Proteins

Jeffrey W. Pollard

MRC Human Genetic Disease Research Group, Department of Biochemistry, Queen Elizabeth College, University of London, Campden Hill, London, England

Introduction

Since O'Farrell (*1*) introduced the improved technique for high resolution two-dimensional polyacrylamide gel electrophoresis (2-D PAGE), it has become one of the most powerful tools for the separation and quantification of proteins from complex mixtures. The principal reason for this is that the method employs separation of denatured proteins according to two different parameters, molecular weight and isoelectric point. Consequently, it has sufficient resolution to separate individual proteins as discrete spots on the gel. Each parameter may also be varied and therefore, with the modification of non-equilibrium pH-gradient electrophoresis (NEPHGE) to analyze basic proteins (*2*), almost any polypeptide may be investigated. Thus to date, the O'Farrell 2-D gel system has no serious rivals, with the possible exception of the Kaltschmidt and Wittmann (*3*) gel system for analyzing ribosomal proteins. Ribosomal proteins, however, may be adequately separated with NEPHGE.

It has been estimated from mRNA complexity studies that individual mammalian cells may contain 10,000 polypeptides ranging in abundance from 10^9 copies/cell to a few hundred. However, by 2-D PAGE analysis only about 1800 individual proteins have so far been detected (*4*), even after long exposures. This is usually interpreted as the inability to detect minor proteins, but Duncan and McConkey (*5*) have argued that in fact 2000 is close to the number of proteins in a cell and that the remaining rare mRNAs are rarely, if ever translated. If this is the case, then 2-D PAGE represents an even more powerful technique than previously expected for investigating changes in cellular physiology. However, these investigations are still limited because, of these two thousand proteins, only a few have been positively identified, although protein maps of individual cell types and sera have been published (*4*). Nevertheless, with subcellular fractionation, antibody detection techniques, and the use of peptide mapping, which has been performed on proteins representing as little as 0.01% of the total protein, it is to be expected that a substantial increase in the number of proteins identified will soon occur, enabling rapid advances in the study of cell biochemistry.

Because of its resolution, the 2-D PAGE technique has been applied to a great number of biological problems ranging from the analysis of proteins in different tissues, under various hormonal states and at different stages of development, to the analysis of cells in culture, and finally to the analysis of polypeptides within a single cell. For these studies of different cellular states, the only technique that may rival the 2-D PAGE method is that of Rot curve analysis of mRNA populations, which in terms of the ability to detect differences (although not in the identification of proteins) has a similar degree of resolution (*5*). Other powerful applications of the 2-D PAGE system include the detection of proteins that contain single amino acid substitutions, which confer a change in isoelectric point on the protein. This has resulted in definitive identification of mis-sense mutations within proteins (*1*) and the visualization of mistranslated proteins (*6*).

The future promises not only further investigations of different metabolic states, but also definitive protein maps, stored as computer matrixes, of cells and sera from a variety of pathological states. These, with the development of specialized equipment to analyze thousands of gels a year in a single laboratory, may enable rapid clinical diagnosis of diseases by comparison of clinical samples to the computerized 2-D PAGE record. Similarly, computer data bases will enable searches for information about the distribution and occurrence of different proteins in hundreds of cell types analyzed at different stages of development, hormonal stimulation, and transformation. However, as yet we await the standardization of these techniques of protein nomenclature, and of rapid, reliable computer analyses (*7*).

Materials

One-Dimensional Isoelectric Focusing

1. Lysis buffer:

   9.5*M* urea
   2% v/v Nonidet P-40 (NP-40)
   2% v/v Ampholines (1.6%, pH range 5–7; 0.4%, pH range 3–10)
   5% v/v β-mercaptoethanol

   This buffer may be stored frozen at −20°C for long periods of time in 0.5 mL aliquots. But do not continually freeze and thaw it. Use an aliquot once and discard the remainder.
2. 30% Acrylamide stock solution:

   28.38% w/v acrylamide
   1.62% w/v *N,N'*-methylene-bisacrylamide

All acrylamide solutions are light sensitive and should be stored in the dark at 4°C. It is usable for at least 1 month. Acrylamide is a potent neurotoxin, so caution should be exercised to protect against contact with the dust. It should *never* be mouth-pipeted.

3. 10% w/v Nonidet in water
4. Anode electrode solution:

   0.01*M* phosphoric acid

   This should be made fresh from a 1*M* stock (63.64 mL phosphoric acid to 1 L water).
5. Cathode electrode solution:

   0.02*M* sodium hydroxide

   This should be made fresh from a 1*M* stock and degassed.
6. Sample overlay buffer:

   8*M* urea
   1% Ampholines (0.8% v/v, pH range 5–7; 0.2% v/v, pH range 3–10)

   This solution may be stored in frozen aliquots.
7. Sodium dodecyl sulfate (SDS) sample buffer:

   0.06*M* Tris-HCl, pH 6.8 at 20°C
   2% w/v SDS
   5% v/v β-mercaptoethanol
   10% v/v glycerol
8. Ampholines are used as supplied in 40% solutions. They should be kept sterile and stored at 4°C.
9. Ammonium persulfate:

   10% w/v solution made up fresh every week and stored at 4°C
10. Pancreatic ribonuclease (50 mg/mL) and deoxyribonuclease (1 mg/mL) in sonication buffer. Store frozen in aliquots.
11. Sonication buffer:

   0.01*M* Tris-HCl, pH 7.4 at 20°C
   0.005*M* $MgCl_2$

Two-Dimensional SDS-Slab Electrophoresis

12. 30% Acrylamide solution:

    29.2% w/v acrylamide
    0.8% *N,N'*-methylene-bisacrylamide

13. Running buffer:

    0.025*M* Tris base
    0.192*M* glycine
    0.1% w/v/ SDS

    For 3 L: 43.2 g glycine, 9.0 g Trisma base, and 3.0 g SDS gives the correct pH; do not titrate this or add any salt.
14. Separating gel buffer:

    1.5*M* Tris-HCl, pH 8.8, at 23°C
    0.4% w/v SDS

15. Stacking gel buffer:

    0.5*M* Tris-HCl, pH 6.8, at 23°C
    0.4% w/v SDS

16. Bromophenol blue:

    0.1% w/v bromophenol blue in water

Solutions 2, 3, 7, 8, 9, 11, 12, 14, and 15 are stored at 4°C, and 4, 5, 13, and 16 at room temperature. All other solutions are stored as indicated.

Method

Sample Preparation

The principle to follow here is to avoid causing any chemical modification of proteins since any charge change will be detected on the gel and will result in an aberrant pattern. The following method should give satisfactory results.

1. Pellets of cells should be resuspended in 100 μL or less of sonication buffer at 4°C. They are sonicated with the microtip of an 150W MSE sonicator for 6 5-s bursts at 8 μm peak to peak. Be very careful not to overheat the sample.
2. Add 2 μL of RNAse and DNAse to 100 μL and incubate for 5 min at 4°C.
3. Add solid urea to bring the sample to 9*M* (1 mg urea/μL of sample), which approximately doubles the volume. Add to this mixture one volume of lysis buffer, take it off the ice, and solubilize the urea with the palm of the hand. Do not overheat since this can result in modifications of the protein. Samples can now be used directly or stored at −70°C until needed.

One-Dimensional Gels (Isoelectric Focusing)

1. Standard glass tubes, 1–1.5 mm internal diameter and 10–15 cm long are thoroughly cleaned in chromic acid (5% Decon may be used, but beware of precipitation).
2. The tubes are rinsed in water, placed in fresh KOH in ethanol (0.4 g KOH to 20 mL ethanol) for 20 min, rinsed thoroughly first with distilled water, then ethanol, and finally allowed to air dry.
3. When dry, the tube bottoms are sealed with three layers of parafilm and lined up vertically around a 250 mL beaker with elastic bands. The tubes are marked to the same point with a felt-tipped pen to ensure that gel lengths are uniform (this is important to facilitate reproducibility between runs).
4. To make 10 mL of isoelectric focusing gel solution add to a 125 mL flask:

Urea	5.5	g
Acrylamide stock (2)	1.33	mL
NP-40 (3)	2.0	mL
Water	1.7	mL
Ampholines (pH 5–7)	0.6	mL
Ampholines (pH 3–10)	0.115	mL

Gels may be made easier to remove by increasing the nonidet concentration by 10%.

5. To make non-equilibrium polyacrylamide gels (NEPHGE), weigh out the same quantity of urea, add the same volume of acrylamide stock and NP-40, but add 2.0 mL of water, 0.25 mL ampholine (pH 7–9), and 0.25 mL ampholine (pH 8–9.5).
6. Dissolve the urea by swirling in a water bath whose temperature is set not higher than 37°C and then briefly degas under vacuum.
7. Add 7 μL (IEF) or 14 μL (NEPHGE) of TEMED and 10 μL (IEF) or 20 μL (NEPHGE) of 10% ammonium persulfate solution. You now have about 15 min to pour the gels at normal room temperature (20°C), but if room temperatures are substantially higher, reduce the TEMED concentration to about two-thirds.
8. Using a long narrow-gage hypodermic needle (or Pasteur pipet) fill the tubes to about 1 cm from the top, being careful to avoid trapping air bubbles (these may be removed by carefully tapping the tubes).
9. Overlay the gel mix with 10 μL of water, allow it to polymerize for 1 h, and then carefully remove parafilm to avoid damage to the bottom of the gel. The parafilm may be replaced by dialysis tubing clipped on with a rubber O-ring to prevent the gel slipping out, but generally this precaution is unnecessary.
10. The gels are placed in the electrophoresis tank (we use an L-shaped tank rather than the conventional round one, since these are easy to manufacture and operate, and can be fitted with locking devices to prevent electric shocks to the operator).
11. The lower chamber is filled with 10 m*M* phosphoric acid (IEF) or 20 m*M* NaOH (NEPHGE), and any trapped air bubbles removed from the end of the gels by a gentle stream of fluid using a bent syringe.
12. For IEF, the water is removed from the top of the gel with a Pasteur pipet and replaced with 10 μL lysis buffer, 10 μL overlay buffer and 20 m*M* NaOH to fill the tubes.
13. The cathode solution is added carefully to the upper chamber and connected to the cathode and the lower chamber to the anode and the gels pre-run at 200 V for 15 min, 300 V for 30 min, and 400 V for 1 h or more. At the end of this procedure, there should not be more than 1.5 mA/5 tubes or 2 mA/12 tubes. Do not cool the

gels during this procedure since the urea will crystallize out.

14. After the pre-run remove the NaOH from the upper chamber carefully so as to avoid contact with the gel surface and discard. Remove the liquid from above the gels and wash the tops with three washes of 20 μL of water.
15. The samples are loaded in a volume of 5–50 μL with a syringe and overlaid with 10 μL sample overlay buffer and 20 m*M* NaOH to fill the tubes.
16. The upper electrophoresis chamber is filled with 20 m*M* NaOH, reconnected to the anode and electrophoresed for 18 h at 400 V. One hour before the termination of the run, turn the voltage up to 1000 V to increase the band sharpness. But do not exceed 10,000 V/h since bands will become distorted.
17. Turn off the power-pack and wait a few seconds, then remove the tubes and force the gels out onto parafilm troughs with a syringe full of water connected to the tubes via a flexible plastic tubing.
18. Put the gels (using the parafilm to handle them) into capped tubes containing 5 mL of sample buffer with 0.002% bromophenol blue. Leave for minimally 20 min without agitation and at this point the gels may be stored at −70°C indefinitely. After defrosting, it is advisable to exchange the sample buffer and leave for a further 30 min before loading onto the second dimension, by which time the second dimension slab gel will have been prepared.
19. For NEPHGE there is no pre-running since a stable pH gradient is not formed; thus samples are loaded directly onto the gels, overlaid with 10 μL of sample overlay buffer and 10 m*M* phosphoric acid to fill the tubes. The upper reservoir is filled with 10 m*M* phosphoric acid and connected to the anode and the gels are run at 400 V for 4.5 h. However, since this is a non-equilibrium procedure, very basic proteins (for example, some ribosomal proteins or histones) may migrate off the end. Thus shorter total voltage-hours need to be used.
20. Investigators may wish to test their procedures at this point and gels may be stained as described in Chapter 6. The bands should be razor sharp. pH gradients

may also be tested by slicing the gel into 0.5 cm pieces, then equilibrated for 1 h in degassed water, and the pH is read using a microelectrode on pH meter. Similarly, a pH contact electrode may be used or alternatively visible pH markers may be bought from a range of suppliers. pH ranges should be from about 3.5 to 7.5 for the IEF gels.

Two-Dimensional SDS–Slab Gel Electrophoresis

1. Assemble the slab gel apparatus (we use the Bio-Rad/Hoeffler design, which gives consistently good results, but any homemade slab gel system is suitable), ensuring that there is a good seal. Make it level and vertical. The plates should have been thoroughly washed in 5% Decon, rinsed with water and ethanol, and air dried. Do not wipe with tissues that leave lint since this will interfere with gel polymerization.
2. The SDS separating gel is made as follows for 100 mL of a 10% gel:

Acrylamide stock (12)	33.3 mL
Water	41.7 mL
Separating gel buffer	25.0 mL

   The solutions are mixed and degassed under vacuum.
3. Fifty μL of TEMED and 333 μL of 10% ammonium persulfate are added and the gel is poured by pipeting the solution down the side of the gel plates to about 2.5 cm from the top. It is overlaid with deionized water applied at one end and left to polymerize for 1 h until the gel interface can be seen as a sharp straight boundary. The gel may be left to stand overnight.
4. The water overlay is removed, the gel surface is washed with water, and the stacking gel poured. This is prepared fresh and to make 25 mL of solution add:

Acrylamide stock (12)	4.0 mL
Water	14.75 mL
Stacking gel buffer	6.25 mL

5. Briefly degas, add 25 μL TEMED and 87.5 μL ammonium persulfate, and pour (taking care to avoid

trapping any air bubbles) up to a level 1 mm below the gel plate edge before inserting a Teflon edge. Overlay with water and allow this solution to polymerize for 30 min.

6. Remove the Teflon strip, rinse the surface of the gel with water, and load the first dimension on top of the second.
7. Take the defrosted, re-equilibrated gel and straighten it in a trough of parafilm. Remove the liquid and apply it directly to the top of the stacking gel. Press down gently with a curved spatula tip from one end to ensure the removal of air bubbles, but be careful not to stretch the gel since this is one stage where variability can be introduced into the procedure.
8. I find the gel adheres; however, the more conventional but slower method is to overlay the gel with 1 mL 1% agarose in SDS sample buffer (not too hot) and then to allow it to set for 5 min. If bromophenol blue had not been included in the sample buffer, add a few drops of 0.1% bromophenol blue over the total length of the gel.
9. The slab gel system is now assembled, the reservoirs filled with running buffer, and any air bubbles trapped under the gel removed with a bent syringe needle or Pasteur pipet. The lower gel reservoir is connected to the anode and the upper reservoir to the cathode, and the gels are then electrophoresed at 9 mA/gel overnight or at 20 mA/gel for 4–5 h until the bromophenol blue reaches the bottom of the gel. It is essential to avoid overheating and thus if the gels are run fast, the whole lot can be run in the cold room or, alternatively, an efficient cooling system may be incorporated into the gel apparatus. After 1–2 h, change the upper reservoir buffer (or alternatively, continuously mix the upper and lower reservoirs with a pump–siphon system) to prevent vertical tailing of the protein spots.
10. At the end of the run, turn off the power, remove the plates, separate them with gentle leverage, and process the gel for staining (Chapters 6 and 13) or autoradiography (Chapters 16 and 17).
11. Briefly, gels may be fixed in 45% (v/v) methanol and 7.5% (v/v) glacial acetic acid (or 15% TCA) and stained

with 0.2% (w/v) Coomassie blue (prepared by dissolving it in a small volume of methanol) in 45% (v/v) methanol and 10% (v/v) glacial acetic acid for 30 min to 1 h. Gels may be destained using several changes in the fixer; the latter may be regenerated by absorbing out the Coomassie blue by passing it through an activated charcoal filter.

12. Gels are then dried down using a gel drier (*see* Chapter 16) and autoradiographed using Kodak-XO-mat-AR film for several different lengths of time. Owing to the saturation of the film, this allows visualization and qualification of a range of proteins with different abundances.
13. Autoradiographs may be quantified using scanning gel densitometry (Chapter 15) or more accurately by the slower method of quantitative roster scanning. However, a better, rapid, and accurate method for handling a large number of proteins is to prepare transparent plastic sheets (the sort used for overhead projector transparencies) as templates. The position of the protein spots on the autoradiogram are marked on the sheet and they are then cut out. This template is used to punch out areas of the gel and these areas are then counted in a scintillation counter. With appropriate labeling the plastic template becomes a permanent record of the protein spot position and may be filed with the resultant counts and autoradiogram. This enables long-term handling of data from a larger number of gels for metabolic experiments. A typical gel pattern is shown in Fig. 1.

Notes

1. The major problems associated with two-dimensional gels are spot streaking and artifactual charge heterogeneity. The principal charge artifacts are produced by post-translational modifications of proteins; for example, the sponaneous deamination of asparagine and glutamine or the oxidation of cysteine to cysteic acid. Care should be taken when preparing extracts and they should always be maintained in the cold and stored in lysis buffer. Proteins may also be carbamylated by

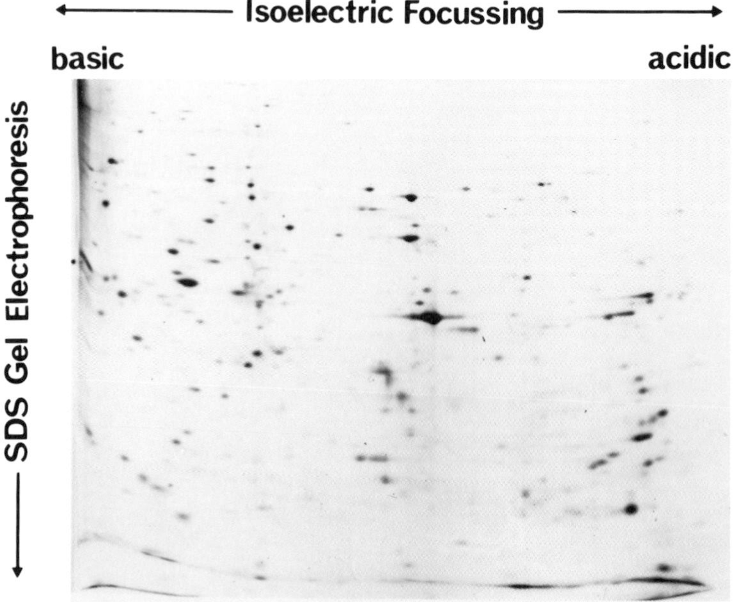

Fig. 1. An autoradiograph of a 10% two-dimensional polyacrylamide gel of ^{35}S-methionine labeled cellular proteins from a temperature-sensitive leucyl-tRNA synthetase mutant of Chinese hamster ovary cells.
About 5 $\times$ 10^5 cpm of ^{35}S-labeled proteins were loaded on the gels that were exposed to the X-ray film for 5 d. Note the characteristic protein spot shape and the distortion at the basic end of the gel caused by edge effects.

isocyanate impurities in the urea solution, and therefore it is worth investing in the highest quality urea, preparing it fresh, and storing it at −20°C. It is also advisable to have ampholines present wherever proteins are in contact with urea and to prerun the isoelectric focusing gels to remove isocyanate contamination. Basic ampholines may precipitate nucleic acids, with the resultant binding of protein producing streaking, but this may be easily solved by treating the sample with ribonuclease and deoxyribonuclease. Nevertheless, sample preparation by the methods described above should result in trouble-free gels.

2. Vertical streaking can also occur and this is caused either by not changing or recycling the running buffer or, more commonly, by poor equilibration of the first-

dimension gels. Occasionally, horizontal streaking can occur owing to poor solubility of proteins, but high concentrations of NP-40 and urea usually increase solubility sufficiently.

3. Spot size increases nonlinearly with high protein concentrations and overloading may cause precipitation at the top of the isoelectic focusing gel, which can streak across the pH gradient. Overloading may also cause pH inversions in the isoelectric focusing dimension. The best separation and resolution is therefore obtained when the lowest amounts of protein are applied to the gel, but concentrations up to 100 μg/gel are tolerated. Obviously, the best results are obtained with small amounts of protein having high specific activities, although the evolution of highly sensitive silver stains may diminish this requirement. If samples are too dilute or at too low a specific activity, they may be precipitated with ammonium sulfate or better, lyophilized, followed by solubilization of the pellet directly in lysis buffer. But remember that any salt will also be concentrated by these techniques and the salt may cause artifacts on the gel. Cells, if they are grown in a small area, for example in a microwell (*see* Chapter 9) may also be taken up directly in lysis buffer and applied directly to the gel, consequently avoiding dilution during sample preparation. Overloading can be particularly a problem with serum because of the enormous concentration of albumin. Methods such as immunoprecipitation of the major protein might improve resolution of minor components. Interestingly, it is possible to polymerize proteins within the isoelectric focusing gel, and this increases capacity, ameliorates the problem of precipitation, and gives acceptable gels.
4. We have found that, when cells in culture are trypsinized, if they are not thoroughly washed, the trypsin may carry over and cause streaking and artifactual gel patterns. Protease degradation, particularly of nuclear proteins, has also proved a problem and this has been remedied by adding protease inhibitors such as sodium bisulfite and diisopropylfluorophosphate. These should be omitted from the 2-D sample gel preparation since they may cause modification of proteins.

5. Ampholines also run as small proteins, are acid precipitable, and will stain. They can be eluted from the gel using the fixative described above and so avoid masking the detection of small proteins migrating near the gel front.
6. Workers often, when they first start running gels, find poor polymerization. This is usually caused by dirty plates or poor quality bisacrylamide or acrylamide. Commercial electrophoresis-grade chemicals are usually of high enough quality, but acrylamide may be purified by heating 1 L of chloroform in a water bath at 50°C and adding to it 70 g acrylamide. Filter through Whatman No. 1 grade filter paper, while it is hot, and leave the resultant liquor on ice for several hours to recrystallize. Collect crystals in a Buchner funnel and wash with cold chloroform. Leave the crystals to dry. Bis-acrylamide may be recrystallized by bringing 1 L of acetone to a boil in a water bath, adding 10 g bis-acrylamide and re-boiling it. Thereafter, continue as described for acrylamide, except that crystalization is performed at −20°C. But remember these chemicals are highly toxic. Finally, to improve polymerization, we make up all solutions in double glass-distilled or deionized water (but be careful of flaking of ion-exchange resin from the column) and we filter all gel solutions through 0.45 μm filters before use.
7. Reproducibility can also be a problem and this is usually caused by batch variation in ampholines. If it can be afforded, batch testing is advisable. Manufacturers have also promised to improve the reproducibility of their ampholines. To date, LKB manufactures the most consistent product. It is also important to keep gel lengths and run times the same from day-to-day and particularly to exercise care in handling the isoelectric focusing gel to avoid stretching it. During equilibration between 5 and 25% of protein may elute from the gel, and variable amounts of protein may also precipitate at the top of the gel during the run. This can be a problem when comparing, for example, time courses of protein synthesis when it is desirable to add constant amounts of protein to the gels. Low concentrations of protein in the sample reduce the precipitation problem and O'Farrell (*1*) has described a technique for rapid equili-

bration involving a high SDS buffer added directly to the isoelectric focusing gel already *in situ* on the slab gel.

8. SDS slab-gel electrophoresis separates according to molecular weight since SDS binds to most proteins on a molar basis (1.4 to 1) giving a uniform charge-to-mass ratio. Thus, dependent on the size range of the proteins to be analyzed, a suitable percentage acrylamide gel may be selected. Gradient gels may also be run and these give greater accuracy to molecular weight determinations of glycoproteins. Nevertheless, anomalous results may be obtained with unusual proteins, for example those containing a large percentage of basic residues, or those with large amounts of carbohydrate. Care should be exercised by the investigator in the interpretation of the experimental results.
9. Given that all these procedures are correctly followed, however, the gels are remarkably reproducible and provide one of the major analytical tools available to protein biochemists.

Acknowledgments

These techniques were derived from the original O'Farrell papers, simplified with the help of J. Parker, J. Friesen, R. Bravo, and J. Celis. This paper was prepared while my work was funded by the MRC (UK), the Cancer Research Campaign and the Central Research Fund of the University of London.

References

1. O'Farrell, P. H. (1975) High resolution two-dimensional electrophoresis of proteins. *J. Biol. Chem.* **250,** 4007–4021.
2. O'Farrell, P. Z., Goodman, H. M., and O'Farrell, P. H. (1977). High resolution two-dimensional electrophoresis of basic as well as acidic proteins. *Cell* **12,** 1133–1142.
3. Kaltschmidt, E., and Wittmann, H. G. (1970) Ribosomal proteins. VII. Two dimensional polyacrylamide gel electrophoresis for fingerprinting of ribosomal proteins. *Ann. Biochem.* **36,** 401–412.

4. Celis, J., and Bravo, R. (eds) (1983) Two dimensional gel electrophoresis of proteins. Academic Press, San Diego, USA.
5. Duncan, R., and McConkey, E. H. (1982). How many proteins are there in a typical mammalian cell? *Clin. Chem.* **28,** 749–755.
6. Pollard, J. W. (1983) Application of two-dimensional polyacrylamide gel electrophoresis to studies of mistranslation in animal and bacterial cells, in *Two-Dimensional Gel Electrophoresis of Proteins* (eds Celis, J., and Bravo, R.), Academic Press, pp. 363-395. San Diego, USA.
7. Anderson, N. G., and Anderson, L. (1982). The human protein index. *Clin. Chem.* **28,** 739–748.

Chapter 11

Starch Gel Electrophoresis of Proteins

Graham B. Divall

Metropolitan Police Forensic Science Laboratory, London, England

Introduction

Starch gel is one of a wide variety of supporting media that can be used for horizontal zone electrophoresis. Such gels are prepared by heating and cooling a quantity of partially hydrolyzed starch in an appropriate buffer solution. The choice of buffer is somewhat empirical and a wide variety of compositions have been used successfully.

A characteristic feature of starch gels is that they exhibit molecular sieving effects. Separation of proteins is achieved, therefore, not only on the basis of differences in charge, but also of differences in molecular size and shape. The contribution of these factors, however, is difficult to control because a starch gel of any particular concentration will contain a range of pore sizes and there is no way of knowing what these are.

The gels may be prepared with total concentrations between 2 and 15% w/v, but 10% is a good starting point for most separations. They are usually cast in rectangular slabs between 1 and 6 mm thick. The 1-mm thick gels (thin layer) are most suitable for the electrophoretic separation of small quantities of protein mixtures, as encountered in forensic analysis. The 6-mm thick gels (thick layer) are used where the availability of sample is not a critical factor. They have been widely used in human genetics to study a variety of protein and enzyme polymorphisms. The preparation of both types of gel is described here.

After electrophoresis, the gels are stained to reveal the separated components. General protein stains can be used, but starch gels are commonly used in conjunction with specific staining procedures in which only one or a group of closely related proteins is located.

Materials

1. Gel and tank buffers are selected empirically to give the best separation for the protein system under investigation. Some compositions that have been used successfully are as follows:

   a. Tank buffer, pH 7.4: 12.11 g Tris; 11.62 g maleic acid dissolved in 1 L water and adjusted to pH 7.4 with 10*N* sodium hydroxide. Gel buffer: tank buffer diluted 1 in 15 with distilled water.
   b. Tank buffer, pH 8.5: 4.41 g sodium barbiturate; 2.34 g diethylbarbituric acid dissolved in 1 L distilled water and adjusted to pH 8.5 with 0.1*N* sodium hydroxide.
   c. Tank buffer, pH 4.9: 28.72 g citric acid; 10.69 g sodium hydroxide dissolved in 1 L distilled water. Gel buffer, pH 5.0: 0.945 g succinic acid; 1.113 g Tris dissolved in 1 L distilled water.
   d. Tank buffer, pH 7.2: 27.21 g boric acid; 1.68 g lithium hydroxide dissolved in 1 L distilled water. Gel buffer, pH 7.4: 0.818 g Tris; 0.378 g citric acid; 0.136 g boric acid dissolved in 1 L distilled water.
   e. Tank buffer, pH 7.9: 18.6 g boric acid; 2 g sodium hydroxide dissolved in 1 distilled water. Gel

buffer, pH 8.6: 9.2 g Tris; 1.05 g citric acid dissolved in 1 L distilled water.

2. Hydrolyzed starch for electrophoresis.
3. Glass plates, 22 cm long, 12 cm wide, and approximately 4 mm thick.
4. Glass edge strips, 22 and 14 cm long, 5 mm wide, and approximately 1 and 3 mm thick.

Methods

Preparation of Starch Gels

Thin Layer Starch Gels

1. Prepare the gel mold by sticking the 1-mm thick glass edge strips to one of the glass plates to give a shallow tray 21 cm long, 14 cm wide, and 1 mm deep. The UV light-sensitive glass adhesives are most suitable for this purpose.
2. For a 10% gel, mix 4 g of the dry starch with 40 mL gel buffer in a 250 mL conical flask to give a lump-free suspension.
3. Heat the suspension over a Bunsen flame while gently swirling the flask. As the temperature rises, the starch begins to dissolve and the solution becomes viscous. It is important to keep the flask moving at this stage since localized heating will cause the starch to burn. Continue the heating and at a certain point the viscosity of the solution will fall very rapidly. Continue to swirl the flask and bring the starch solution to a boil. Continue to boil for a few seconds. The bubbles will be seen to change from a frothy opalescent appearance to a clear transparent form.
4. Degas the solution by applying a vacuum to the flask for approximately 5 s. The solution will boil again under the reduced pressure. Release the vacuum slowly or else the hot starch solution will be violently mixed and can spit out of the flask.
5. Place the mold on a horizontal surface and pour the hot starch solution across the width of the plate at one end. Spread the solution across the plate to form an even layer. This is best achieved by using a bevelled Perspex

scraper placed across the width of the plate and drawn down its length, allowing the excess starch solution to flow off the end.

6. Leave the plate for 5 min to allow the starch solution to cool and gel. Transfer the gel plate to a moisture cabinet at 4°C.

Thick Layer Starch Gels

1. Prepare the gel mold by sticking the 3-mm thick glass edge strips to one of the glass plates. Place a second set of edge strips on top of the first. These are held in place by smearing the joining surfaces with a non-silicone grease. This will give a tray 21 cm long, 14 cm wide, and 6 mm deep.
2. For a 10% gel, mix 20 g of the dry starch with 200 mL gel buffer in a 2 L conical flask to give a lump-free suspension.
3. Heat the suspension with continuous vigorous swirling and bring the solution to a boil. Continue boiling for a few seconds and then degas the solution by applying a vacuum to the flask.
4. Place the mold on a horizontal surface, pour the hot starch solution into the center, and allow it to spread over the surface of the mold. If any bubbles form or if the solution does not reach the edge, touch the solution at the point with a rubber-gloved finger.
5. Leave the plate for 30 min to allow the starch solution to cool and gel. It is important that the plate not be disturbed during this period or the gelling solution will ripple. Transfer the gel plate to a moisture cabinet at 4°C.

Sample Application

Thin Layer Gels

1. Make sample application slots in the gel across the width of the plate approximately 6 cm from the cathode end. This is most easily achieved by using a comb made of thin (0.5 mm) plastic, with teeth 1 cm wide and 0.5 cm apart. Press the comb into the gel and remove it cleanly.

2. Soak pieces of 1-cm long cotton thread (pulled from a plain cotton bed sheet) in the sample solution and insert into the sample slots. Each thread absorbs approximately 2–3 μL of sample solution. Ensure the threads are pressed well down to touch the base glass plate. The threads are best held and manipulated using a pair of fine forceps.

Thick Layer Gels

1. Make sample application slots in the gel across the width of the plate approximately 6 cm from the cathode end. These can be made individually by inserting a piece of razor blade, approximately 1 cm wide, vertically into the gel or using a comb as described for the thin layer gels.
2. Soak a piece of Whatman 3 mm filter paper, approximately 1 × 0.5 cm, in the sample solution and insert into a sample slot. Ensure the inserts reach the base glass plate. The inserts absorb approximately 15–20 μL of sample solution.

Electrophoresis

1. Place the gel plate in an electrophoresis tank and apply wicks to both ends of the gel. The wicks, consisting of several thicknesses of filter paper or pieces of sheet sponge and cut to the width of the gel, should be soaked in tank buffer, and overlap the gel surface by 1–2 cm. They can be held in place by a plain glass plate, 22 × 15 cm, which will also cover the gel and reduce evaporation.
2. The gels must be cooled during electrophoresis by placing the whole tank in a cold room or refrigerator at 4°C, or by using a cooling plate through which water is circulated at 4–8°C.
3. It is difficult to generalize about the conditions of electrophoresis: time, voltage, and current are selected to give the best separation for the protein system being studied. As a starting point, however, potentials between 2 and 20 V/cm are used with running times between 2 and 16 h.

Fig. 1. Part of a thin-layer starch gel electrophoresis plate used to detect phenotypic variation in human red cell phosphoglucomutase (PGM). The samples were approximately 2–3 μL of human red cell lysate applied to the gel on 1-cm cotton thread inserts. Electrophoresis was performed at 6 V/cm for 16 h in a cold room at 4°C. The plate was then developed for PGM activity using a zymogram agar overlay (*see also* Chapter 12 for the same analysis by isoelectric focusing).

4. After electrophoresis, remove the top plate and wicks, lift the gel plate from the electrophoresis tank, and apply a suitable stain. A typical electrophoretic run is shown in Fig. 1. The stain may be applied to the top surface of the thin layer gel, but the thicker gels must

be sliced. To do this, remove the filter paper inserts and the second set of edge strips. Place a glass plate on the surface of the gel and draw a thin wire or cutting blade through the gel. During the cutting process, ensure that the wire or blade is kept firmly pressed to the surfaces of the bottom edge strips. Remove the top glass plate and roll a piece of damp filter paper onto the gel surface. The upper half of the gel will adhere to the paper and can be peeled away from the lower half. The cut surfaces of each half can now be stained.

Notes

1. A wide variety of buffers have been successfully used. The pH, ionic strength, and buffer constituents are all important considerations when trying to achieve a particular separation. When studying enzymes, start with a buffer whose pH is near the optimum and use short running periods with low voltages, e.g., 2–4 h at 2–4 V/cm. It is sometimes necessary to incorporate known activators or cofactors into the buffers, e.g., Mg^{2+} ions or EDTA, since this can improve the resolution into distinct zones and increase the staining intensities of enzymes.
2. Hydrolyzed starch suitable for electrophoresis can be prepared in the laboratory by warming granular potato starch in acidified acetone. Most laboratories, however, use hydrolyzed starch powder that is commercially available. The methods described here have assumed 10 g/100 mL to give a 10% gel, but the exact amount of powdered starch required varies from batch to batch and is always given with the manufacturers instructions.
3. The starch solutions take between 5 and 30 min to gel, but it has been noticed on numerous occasions that superior resolution of protein bands is achieved if the gels are stored in a moisture cabinet overnight.
4. After preparation, the gels should appear clear or slightly opalescent and homogenous. The presence of undissolved starch, which appears as white speckles, bubbles, bacterial growth, or any other artefact renders the gel unsuitable.

5. The sensitivity of the method is ultimately dependent on the staining procedure used. Samples that have a total protein concentration of between 1 and 100 mg/mL are generally used.

References

1. Smithies, O. (1955) Zone electrophoresis in starch gels: group variations in the serum proteins of normal adults. *Biochem. J.* **61,** 629–641.
2. Wraxall, B. G. D., and Culliford, B. J. (1968) A thin-layer starch gel method for enzyme typing of bloodstains. *J. Forensic Sci. Soc.* **8,** 81–82.

Chapter 12

Isoelectric Focusing in Ultrathin Polyacrylamide Gels

Graham B. Divall

Metropolitan Police Forensic Science Laboratory, London, England

Introduction

Isoelectric focusing is an electrophoretic method for the separation of amphoteric macromolecules according to their isoelectric points (p*I*) in a stabilized natural pH gradient. It is a sensitive and reproducible technique, particularly valuable in the separation of closely related proteins that may not be easily separated by other techniques.

The method consists of casting a thin layer of polyacrylamide gel containing a large series of carrier ampholytes. A potential difference is applied across the gel and the carrier ampholytes arrange themselves in order of increasing p*I* from the anode to the cathode. Each carrier ampholyte maintains a local pH corresponding to its p*I*, thus creating a uniform pH gradient across the plate. Samples are applied to the gel surface and under the influence of the electric field each individual protein component migrates to the region in the gradient where the pH corresponds to its isoelectric point. At its p*I*, a protein is electrically neutral and it becomes stationary or focused at that point in the gel.

When focusing is complete, the separated components are detected by applying a general or specific stain to the gel surface.

Materials

1. Acrylamide, *N,N'*-methylene bisacrylamide and analytical grade sucrose.
2. Stock riboflavin solution, 1 mg in 10 mL distilled water. Store the solution in a brown glass bottle at 4°C.
3. Carrier ampholytes of the desired pH range. The author has experience only with those available from LKB, Sweden (Ampholine[R]). They are supplied as 40% w/v solutions. Store at 4°C.
4. Anolyte and catholyte wick solutions consisting of 1% v/v acetic acid and ethanolamine, respectively.
5. Plain glass plates, 22 cm long, 15 cm wide, and approximately 4 mm thick.
6. Fixing solution: 5 g sulfosalicylic acid and 10 mL trichloroacetic acid in 90 mL distilled water.
7. Destaining solution: 300 mL methanol, 100 mL glacial acetic acid, and 600 mL distilled water.
8. Staining solution: 0.2 g Coomassie Brilliant Blue R.250 in 100 mL destaining solution. The solution should be filtered before use.

Method

1. Prepare the ultrathin mold by sticking strips of PVC electric insulating tape, 1 cm wide and approximately 0.15 mm thick, around the edge of a clean glass plate. This will give a very shallow tray, 18 cm long, 13 cm wide and 0.15 mm deep. Avoid overstretching the tape since it will later shrink and gaps will occur at the corners. Small gaps are tolerable, but larger ones (1

mm or more) will hinder polymerization of the acrylamide and cause an unevenness in the gel at these points.

2. Prepare the following solution, which is sufficient for making 12 plates: 3.88 g acrylamide, 0.12 g *N,N'*-methylene bisacrylamide, 10 g sucrose, and 80 mL distilled water. Ensure that all of the components are completely dissolved. This is best done with a magnetic stirrer, but vigorous mixing of the solution should be avoided. The solution should not be stored.
3. Add 0.6 mL riboflavin solution. The riboflavin does not completely dissolve in the stock solution, so that this solid material should not be picked up.
4. Add 4 mL of carrier ampholyte solution of the appropriate pH range. The final concentration of carrier ampholytes is approximately 2% w/v. Ensure all the components are uniformly mixed by gently swirling the flask. Degassing has been found to be unnecessary so long as violent mixing has been avoided.
5. Place the glass mold in a spillage tray (photographers' developing trays are ideal for this purpose) with the taped surface uppermost. Wipe the surface with an alcohol moistened tissue to remove any traces of grease.
6. Using a disposable plastic syringe, transfer approximately 7 mL of the acrylamide–ampholyte solution to along one of the short edges of the glass mold.
7. Take a plain glass top plate (20 × 15 cm), wipe the surfaces with alcohol, and place one of the short edges along the taped edge of the mold adjacent to the acrylamide–ampholyte solution. Gradually lower the top plate and allow the solution to spread over the mold. Care must be taken to avoid trapping air bubbles between the two glass plates. If this happens, raise the top plate to remove the bubble, then lower it again. Excess solution will flow into the spillage tray. Press the top plate into firm contact with the taped edge of the bottom plate.
8. Lift the complete mold and top plate out of the spillage tray. Gently scrape the bottom of the mold on the edge of the tray to remove excess solution.
9. Place the whole plate in UV light for at least 3 h to allow photopolymerization to occur. A viewing table il-

luminated by white strip lights or bright direct sunlight provides sufficient UV light. To save space plates may be stacked, but no more than three in a pile, and each plate must be separated from the next by glass strips to prevent the plates sticking to each other.

10. Pour the excess acrylamide–ampholyte solution from the spillage tray into a clean flask. The dregs from four plates will be sufficient to pour one further plate.
11. Remove the plates from the UV light source and wipe the outside surfaces with a wet tissue to remove any solid material. This is important since the whole surface of the bottom plate must be in intimate contact with the cooling plate during the focusing run.
12. Store the plates in the dark at 4°C. They can be kept for several months, but should not be used if the gel has discolored or dried. If the plates are used immediately, it is still advisable to cool them to 4°C since this facilitates removal of the top plate.
13. Remove the top plate immediately prior to sample application by placing the whole plate on a horizontal surface and inserting a scapel blade between the top and bottom plate in one of the corners. Carefully twist the blade and pry the top plate away. Sometimes the gel sticks to the top plate, but as long as it remains in contact, the top plate plus gel may be used for focusing. Occasionally, part of the gel will stick to the top plate and part to the bottom plate. The whole lot will then have to be discarded.
14. Absorb sample solutions on 5 × 5 mm pieces of Whatman No. 1 filter paper and lay these on the gel surface down the length of the plate at approximately 5 cm from the anodal edge. Each piece of filter paper absorbs approximately 5–8 μL of sample solution. Samples that have a total protein concentration of between 0.05 and 10 mg/L are generally used.
15. Place the gel plate on a cooling plate through which water is circulating at approximately 4–8°C.
16. Apply electrode wicks to the anode and cathode edges of the gel. These consist of 18 × 1 cm strips of Whatman No. 17 filter paper saturated with 1% acetic acid (anolyte) and 1% ethanolamine (catholyte), respectively. Uniform soaking of the wicks is important in preventing wavy iso-pH bands.

17. Place a platinum electrode on each wick. These can consist of an 18 cm-long strand of platinum wire stretched and held taught across an 18 × 1 cm piece of Perspex. Hold the electrodes in place with a glass top plate.
18. Apply a potential difference of between 130 and 160 V/cm across the plate. This represents (for a 12 cm electrode gap) a working potential of approximately 1600–2000 V. A special high voltage power pack will, therefore, be required. Starting currents are usually between 5 and 8 mA.
19. Focusing takes between 3 and 6 h. This can be checked by placing identical samples on the cathode and anode sides of the gel. Equilibrium focusing conditions have been reached when the same components in each sample are seen to coincide.
20. When focusing is complete, disconnect the power supply, remove the top plate, electrodes, and wicks, lift the gel plate from the tank, and apply a general or specific stain to the gel surface. A typical example is shown in Fig. 1.
21. Many general protein stains form insoluble complexes with the carrier ampholytes and these must be removed from the gel before the stain is applied. A suggested procedure is as follows: Place the gel plate in the fixing solution for 20 min; transfer the plate into destaining solution for 15 min; transfer into staining solution for 30 min; then finally into destaining solution for 20 min. All stages are performed at room temperature.

Notes

1. When making ultrathin gels, some workers recommend employing the process of silanization (in which the polyacrylamide gel is chemically bonded to the glass plate, thus preventing the gel from becoming damaged or torn during use). In practice however, we have found this an unnecessary step.
2. Prefocusing of the plates prior to sample application, although carried out by some workers, is an unnecessary step in most instances.

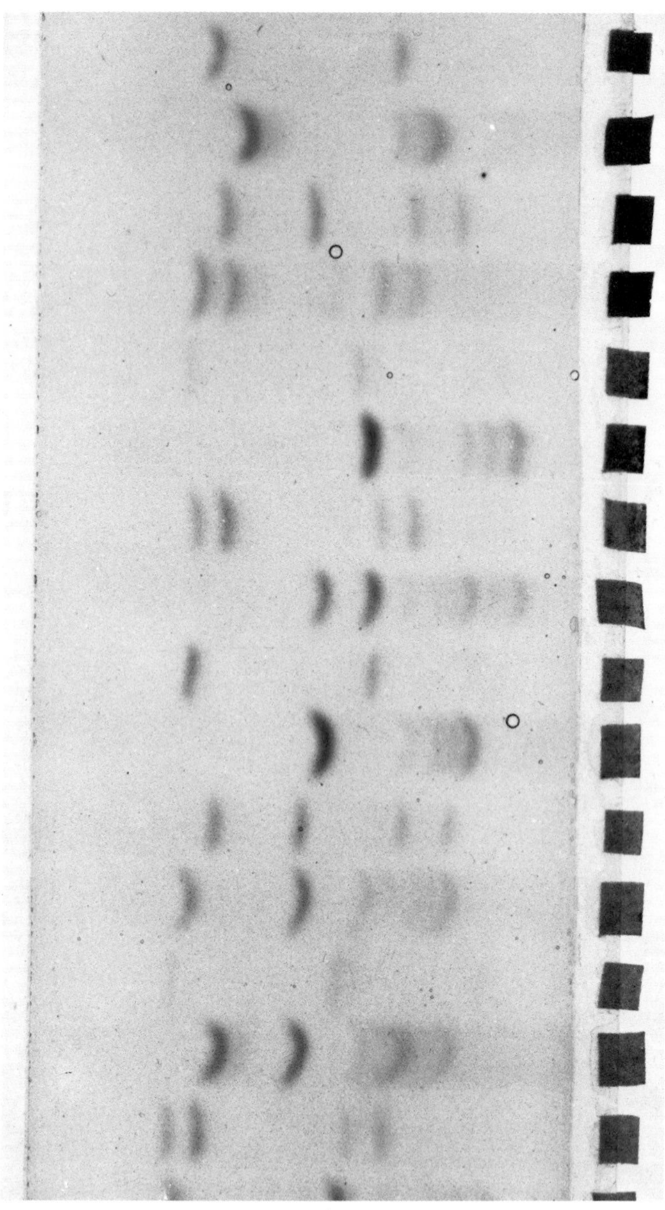

Fig. 1. Part of an ultrathin IEF plate used to detect phenotypic variation in human red cell phosphoglucomutase. The samples were approximately 5–8 μL of human red cell lysate (1:5 with distilled water) applied to the gel surface on filter paper squares at 5 cm from the anode edge. They were subjected to electrophoresis at 140 V/cm for 3 h in a pH 5–7 gradient. The run was stopped before focusing was complete and the plate was developed for PGM activity using a zymogram agar overlay.

3. Theoretically, it should make no difference where the samples are applied since all components will move in the pH gradient to positions corresponding to their isoelectric points. In reality, this is not always the case. Cathodal application may be more appropriate, for example, if a particular protein is easily denatured at acid pH values.
4. The electrode wick solutions (acetic acid and ethanolamine) have been used successfully for separations in the following gradients: pH 3–10, pH 5–7, pH 4–8, and pH 4–6.
5. Adequate cooling is essential, but remember that at low temperatures proteins take longer to focus. For some systems, overcooling produces an inferior resolution of the components.
6. Considerable advantages can be obtained by not allowing the proteins to reach their isoelectric points. The runs are shorter, resolution can be improved, and an overall increase in clarity is sometimes observed. Under such conditions, the method is really one of electrophoresis in a pH gradient rather than true isoelectric focusing.

References

1. Randall, T., Harland, W. A., and Thorpe, J. W. ‹1980) A method of phenotyping erythrocyte acid phosphatase by isoelectric focusing. *Med. Sci. Law* **20,** 43–47.
2. Divall, G. B. (1981) Studies on the use of isoelectric focusing as a method of phenotyping erythrocyte acid phosphatase. *Forensic Sci. Int.* **18,** 67–78.
3. Divall, G. B., and Ismail, M. (1983) Studies and observations on the use of isoelectric focusing in ultra-thin polyacrylamide gels as a method of typing human red cell phosphoglucomutase. *Forensic Sci. Int.* **22,**253–263.

Chapter 13

High-Sensitivity Silver Staining of Proteins Following Polyacrylamide Gel Electrophoresis

Keith Gooderham

MRC Clinical and Population Cytogenetics Unit, Western General Hospital, Crewe Road, Edinburgh, United Kingdom

Introduction

Polyacrylamide gel electrophoresis is a simple, inexpensive, yet highly versatile and powerful method for the analysis of complex mixtures of proteins. In part, the success of this method has resulted from the ease with which the fractionated proteins can be detected with the blue dye, Coomassie brilliant blue R 250. However, though this stain has proved to be ideal for many of the more traditional applications of this method, it is of limited sensitivity. In particular, the recent development of two-dimensional gel electrophoresis (Chapter 10) and *in situ* peptide mapping techniques (Chapter 22) have demanded increasingly more sensitive detection methods. In part, this requirement has been met by the use of radioactively labeled proteins, followed by either autoradiography or fluorography (Chapter 9). The principal drawback of this approach is that it involves additional sample preparation and it is often difficult to label the proteins to a sufficiently high specific activity, particularly where proteins are obtained from dilute physiological samples, and so on.

The recent development of a number of very sensitive silver stains, which are between 20 and 200 times more sensitive than Coomassie brilliant blue (*see,* for example, Fig. 1 of Chapter 22), has been a very valuable addition to the range of protein detection methods that can be used with polyacrylamide gel electrophoresis. These silver staining procedures have been developed from a variety of histochemical and photochemical techniques, where silver has long been recognized as a valuable and sensitive reagent. Silver staining has the considerable advantage of not requiring any special sample preparation and it can also be used in conjunction with standard Coomassie stains. A number of problems arise from the novelty of this method and our limited understanding of the precise chemistry of the reactions involved. The most significant of these are: lack of reproducibility, nonlinear staining, preferential staining of different proteins, and nonspecific background staining. However, silver staining remains a valuable method for the detection of proteins in polyacrylamide gels.

The procedure described here is based on a method originally published by Oakley et al. (*1*). Although this method is neither the most sensitive or rapid of the methods that are currently available, we have found that it is both reliable and easy to use.

Materials

With the exception of Solution C, all of the following solutions are prepared from analytical grade reagents. Deionized, glass-distilled water is used for the preparation of all solutions. The volumes of these solutions are sufficient to stain a gel measuring approximately 15 by 15 cm; where either larger or smaller gels are to be stained, these volumes should be adjusted accordingly.

1. *Solution A.* 500 mL of 50% (v/v) methanol, 10% (v/v) acetic acid.
2. *Solution B.* 2000 mL of 20% (v/v) methanol, 7% (v/v) acetic acid.
3. *Solution C.* 500 mL of 10% (v/v) glutaraldehyde is prepared by diluting 100 mL of 50% glutaraldehyde solution (laboratory reagent grade) to 500 mL with water.
4. *Solution D.* 200 mL of ammoniacal silver nitrate solution is prepared as follows: to approximately 170 mL of water, add 2.8 mL of concentrated ammonia solution (0.88 sp.gr.) and 3.75 mL of 1*N* NaOH solution. A volume of 8 mL of 20% (w/v) silver nitrate solution is then slowly added to this solution while vigorously mixing the solution. The solution is then made up to 200 mL with water.
5. *Solution E.* 1000 mL of 0.005% (w/v) citric acid, 0.019% (v/v) formaldehyde [0.5 mL of 38% (v/v) formaldehyde per liter].
   Solutions A and B can be kept for up to 2 wk before being used. All other solutions are prepared as required and used immediately. Both the concentrated ammonia and glutaraldehyde solutions should be stored at 4°C when not in use.
6. The only items of specialized equipment required for the following protocol are an orbital shaker table and a number of *clean* glass "gel dishes"; shallow baking dishes are ideal for this purpose.

Method

In order to prevent the surface of the gel becoming contaminated by finger marks and so on (which are readily detected by this method), it is essential that talc-free plastic gloves are worn whenever the gel is handled. However, in the majority of the following steps, the various solutions can be removed with the aid of a water pump, thus avoiding any direct contact with the gel.

With the exception of Solution D (Step 5) where only 200 mL of this reagent is used, approximately 500 mL of each solution is used for each wash and the gel is continually agitated at between 20 and 50 rpm on an orbital shaker table.

1. Proteins are separated by polyacrylamide gel electrophoresis (*see* Chapters 6–10) and the gel is then placed in a glass dish and washed for 2–16 h in Solution A.
2. The gel is then washed for a further 2–16 h in Solution B.
3. This solution is then replaced with Solution C and the gel is washed for a further 30 min, after which time it is washed with at least four changes of distilled water for a minimum of 2 h and preferably overnight.
4. Upon completing the wash step, the ammoniacal silver nitrate solution (Solution D) is poured into the dish and the gel is vigorously agitated (~200 rpm) for 15 min. The gel is then briefly rinsed with two changes of distilled water and transferred to a clean gel dish between the first and second rinses.
5. The proteins are then visualized by the addition of Solution D and gently rocking the gel dish. When the image has developed to the required intensity (usually within 5 min), the gel is again washed in Solution B for about 10 min before finally being stored in this solution in the dark at 4°C. Although the silver-stained proteins remain clearly visible for at least 18 h, there may be some slight change in color as well as an increase in the nonspecific background staining during this time and the gel should therefore be photographed as soon as possible after staining.
6. During the above staining procedure the gel will undergo a considerable increase in size and the proteins frequently appear as diffuse bands. This effect can be reversed by leaving the gel for 2–16 h in Solution A (in the dark). The high concentration of methanol in this solution will dehydrate and shrink the gel and the proteins will then appear as intense, sharp bands.

Notes

1. The importance of wearing talc-free plastic gloves and of touching the gel as little as possible has already been stressed. In addition it is important that all of the glassware used for the preparation of the gel, as well as the silver staining solutions, is scrupulously clean.
2. Ammoniacal silver nitrate solution (Solution D) is potentially explosive when dry, and therefore should be collected after it has been used for staining the gel and precipitated by the addition of an equal volume of 1*N* hydrochloric acid. When large numbers of gels are being silver stained, sufficient silver chloride may be collected to make the recovery of the silver worthwhile. However, if this method is used only occasionally, the silver chloride can be washed down a drain with a large volume of cold water.
3. In gels where either the proteins have been stained too heavily or there is too much background staining, the staining can be reversed with Kodak rapid fixer, as described by Oakley et al. (*1*).
4. If gels are to be pre-stained with Coomassie brilliant blue, gels are treated with a 0.5% (w/v) solution of Coomassie brilliant blue R 250 in Solution A for 2–18 h, followed by destaining against several changes of Solution B. The gel can then be silver stained by starting either at step 3 (*see* Method) or the gel can be "totally" destained by washing for 1–2 d in Solution A and then continuing from step 2 (*see* Method). This "total" destaining step will remove the Coomassie stain from all but the most intensely stained proteins and will ensure a more uniform deposition of silver.
5. The above procedure is designed for slab gels that are up to 1 mm thick. Where thicker gels are to be stained this method may fail to produce satisfactory results, with only the proteins at the surface of the gel being stained. Marshall and Latner (2) have recently published a technique designed for staining thick gels, and this should be used when the method described here proves unsatisfactory.
6. Most radioactively labeled proteins (^{125}I, ^{32}P, ^{35}S, and ^{14}C) can be detected by either autoradiography or

fluorography, as appropriate, following silver staining. However, gels containing proteins that have been labeled with tritium are unsuitable for fluorography following silver staining because the silver grains absorb most of the weak beta emissions from this isotope.

References

1. Oakley, B. R., Kirsch, D. R., and Morris, N. R. (1980) A simplified ultrasensitive silver stain for detecting proteins in polyacrylamide gels. *Anal. Biochem.* **105,** 361–363.
2. Marshall, T., and Latner, A. L. (1981) Incorporation of methylamine in an ultrasensitive silver stain for detecting protein in thick polyacrylamide gels. *Electrophoresis* **2,** 228–235.

Chapter 14

Quantification of Proteins on Polyacrylamide Gels (Nonradioactive)

B. J. Smith

Institute of Cancer Research, Royal Cancer Hospital, Chester Beatty Laboratories, London, England

Introduction

It is frequently necessary in biochemical experiments to quantify proteins. There are various methods for estimation of the concentration of total protein in a sample, such as total amino acid analysis, the Biuret reaction, and the Lowry method [*see* ref. (*1*)], but these do not allow quantification of one protein in a mixture of several. This may be done by chromatography and estimation of the content of the appropriate peak in the elution profile by virtue of its absorption of light at, say, 220 nm. However, it is usually quicker, easier, and more economical to quantify proteins that have been separated from each other on a polyacrylamide gel. This is done by scanning the gel and by densitometry of the stained bands on it. Microgram quantities of protein may be quantified in this way. The method described below for quantitative staining uses Procion Navy MXRB and is suitable for proteins on acid/urea or SDS polyacrylamide gels (*see* Chapters 6 to 8).

Materials

1. A suitable densitometer, e.g., the Gilford 250 spectrophotometer with scanning capability and chart recorder, or the Joyce-Loebl "Chromoscan."
2. Protein stain.

   Procion Navy MXRB 0.4 g

   Dissolve in 100 mL methanol, then add 20 mL glacial acetic acid and 80 mL distilled water. Make fresh each time.

3. Destaining solution.

Methanol	100 mL
Glacial acetic water	100 mL
Distilled water	800 mL

Method

1. At the end of electrophoresis, immerse the gel in Procion Navy stain, and gently agitate until the dye has fully penetrated the gel. This time varies with the gel type [e.g., 1.5 h for a 0.5 mm-thick SDS polyacrylamide (15%T) gel slab], but cannot really be overdone.
2. At the end of the staining period, decolorize the background by immersion in destain, with agitation and a change of destain whenever the destain becomes deeply colored. This passive destaining may take 24–48 h even for a 0.5 mm-thick gel. Thicker gels will take longer.
3. Measure the degree of dye binding by each band of protein by scanning densitometry of the gel at a wavelength of 580 nm. The total absorption by the dye

in each band (proportional to the area of the peak in the scan profile) may be automatically calculated by integration of peaks in the scan profile, but if not then the peaks in the chart recording may be cut out and weighed. For standard curves, the absorption at 580 nm (i.e., peak size) is plotted versus the weight of protein in the various samples.

Notes

1. For quantification of proteins on gels, three simple conditions need to be fulfilled:
   (i) Protein bands on the gel should be satisfactorily resolved.
   (ii) The dye used should bind to the protein of interest.
   (iii) Since one sample will be compared to others, to overcome errors caused by differences in sample sizes, the binding of the dye should be constantly proportional to the amount of protein present over a suitably wide range.
2. Errors that are difficult to eradicate arise when the above condition (i) is not fulfilled. Ideally, another electrophoresis system that *does* give sufficient resolution should be used, but otherwise the operator has to decide where the division between two overlapping peaks should be and where the baseline is. Use of a narrower beam of light for densitometry will improve resolution, but may worsen the baseline because of the increased effects of bubbles, dust, and so on.
3. Conditions (ii) and (iii) should be checked by construction of a standard curve for the protein(s) of interest. The range of protein quantities used should cover that to be found in experiments, and for accurate results the plot of dye-binding (measured by densitometry at a particular wavelength) versus protein quantity should be linear (or approach close to it) over that range.
4. Any stain may be used for quantification provided that it meets the above requirements (ii) and (iii). However, various factors may influence the choice of dye to be used. This can be illustrated by comparing Coomassie

Brilliant Blue R250 (BBR 250) with Procion Navy MXRB. Firstly, given time, Procion Navy can penetrate dense bands of protein and stain them, but Coomassie BBR 250 seems unable always to do this, so that the size of band that the protein forms may affect its staining. Thus, since the size of a protein band will generally increase with migration during electrophoresis, the extent of electrophoresis may have a small effect on staining by this dye. Secondly, when Coomassie BBR 250 binds to a protein, the resulting complex may be colored blue or red or anything in between, depending on the chemical structure of the protein. Such production of a variety of colors upon complexing of dye with proteins is called the "metachromatic effect." Maximum sensitivity is achieved in this technique by densitometry at the wavelength at which the dye–protein complex absorbs maximally. The metachromatic effect, therefore, dictates multiple estimations at wavelengths optimal for each complex, or gives suboptimal sensitivity. Procion Navy gives the same color with every protein. Thirdly, Coomassie BBR 250 binds proteins by electrostatic forces and can be completely removed from protein by extensive destaining. Thus, destaining may introduce an element of variability to the experiment. Results are more consistent with Procion Navy, which binds covalently to proteins so that prolonged destaining does not remove it. Fourthly, the efficacy of Coomassie BBR 250 varies from batch to batch of the solid dye and also with the method of its dissolution in making the stain (*see* Chapter 6). We have not seen this with Procion Navy MXRB. From the above it may be seen that Procion Navy MXRB (and probably other members of the Procion dye family) is potentially the better choice for quantification purposes. For the proteins studied in our own laboratory, at least, it also gives stoichiometric binding to proteins over a wider range than does Coomassie BBR 250 and so it is the dye of choice for quantification, even though it is severalfold less sensitive than Coomassie BBR 250.

5. Standard curves of Procion Navy MXRB binding to various proteins are straight lines passing through the

origin, although they do plateau at higher protein quantities. The upper limit is dependent on the size of the gel, but in a 0.5 mm-thick gel the standard curve was found to be linear beyond a loading of 6 μg of histone H1, and up to about 30 μg for a mixture of four proteins (histones) together (2). On such gels, a loading of less than 0.5 μg is detectable and 1 μg is sufficient for quantification.

6. The whole process of quantification should be standardized to reduce variation. Sources of error should be recognized and countered if results are to be accurate. Thus, the densitometer should give a linear response over the range of dye densities measured. Sampling errors are inevitable, but their effect can be reduced by repetition and averaging the results. The process of electrophoresis should be standardized, using samples of the same volume and electrophoresing for the same time at the same voltage in each experiment. Multisample slab gels are ideal for this purpose. The staining time should also be the same each time. Since variation in band widths will still occur, in a slab gel at least, and because irregularities in band shape will occur occasionally in any case, the whole of every stained band should be quantified. If the band width is greater than the width of the densitometer's light beam, and the raster type of scan is not available, the whole gel may be sufficiently reduced in size by equilibration in aqueous ethanol. Too much ethanol causes the whole polyacrylamide gel to become opaque, but if this happens the gel may be rehydrated in a lower-percent ethanol solution. For shrinking 15% polyacrylamide gels, 40% v/v ethanol in water for 1 h or so is suitable, reducing their size by about a third. If the sample width is still too great, it can be divided up and each section estimated separately. These estimates are then summed for an estimate of the whole sample.

   Despite efforts to reduce variations between experiments, they may still occur. In this case it may be necessary to construct a standard curve for each experiment. The type of standard curve required differs according to the aim of the experiment, which may be of the following types:

(a) Simple comparison: determination of the relative concentrations of the same protein in two or more samples.
(b) Complex comparison: determination of the relative concentrations of two or more proteins in the same or different samples.
(c) Absolute determination: determination of the amount of a protein in one or more samples (that is, on a weight or molar basis).

In experiments of type (a) the dye-binding capabilities of the (identical) protein in the different samples will be the same and all that is needed is to know that dye binding is stoichiometric over the appropriate range of quantities. Thus, the standard curve can be set up with a serially diluted (or concentrated) sample, without any accurate protein quantification involved. To compensate for sampling errors, it is often useful to relate the protein of interest to an "internal standard" of another protein that is known to be at the same concentration in all samples. This internal standard will indicate the magnitude of difference in size between samples on the gel, and this factor can then be used to adjust the results of the experimental protein. It must be remembered that the internal standard protein should also have stoichiometric dye binding capabilities over the appropriate range.

Two different proteins will not necessarily have the same dye-binding capabilities, so if they are to be compared on a weight or molar basis [experiment type (b)] it is necessary to determine the characteristics of each. Thus, the pure proteins are isolated, and each is accurately quantified (say, by amino acid analysis) and used to construct a standard curve. Experimental samples can be compared with these standard curves to give accurate weight measurements or it may be sufficient to determine the ratio between dye-binding properties of the different proteins for subsequent use with only one of the standard curves.

Experiments of type (c) are similar to those of type (b) in that accurately constructed standard curves are required using well-quantified protein standards. Note that simple weighing is usually not accurate enough for such work as this, for it cannot distinguish between protein and nonproteinaceous constituents (such as contaminating dust).

For accurate results with experiment types (b) and (c), standard curves need to be run in each experiment (even if they are relatively crude, with only 4 or 5 points).

7. Destaining of Procion Navy-stained gels may be speeded up by driving free dye out of the gels by electrophoresis. The protein–dye complex remains in the gel. To do this, immerse the gel in destaining solution. Place a platinum electrode on either side of the gel and apply a dc electric field across the gel. Change the buffer frequently (as it becomes colored) or cycle the buffer through a decolorizing agent (i.e., an agent that binds the dye). Disposable paper tissues have proven satisfactory as a cheap, disposable decolorizing agent.

References

1. Scopes, R. K. (1982) *Protein Purification. Principles and practice.* Springer-Verlag, New York.
2. Smith, B. J., Toogood, C., and Johns, E. W. (1980) Quantitative staining of submicrogram amounts of histone and high-mobility group proteins on sodium dodecylsulphate-polyacrylamide gels. *J. Chromatog.* **200,** 200–205.

Chapter 15

Computer Analysis of Gel Scans

Harry R. Matthews

University of California, Department of Biological Chemistry, School of Medicine, Davis, California

Introduction

Gels are frequently scanned either to obtain quantitative data from the patterns of stain or radioactivity or to facilitate comparisons between samples. Both these processes can be greatly enhanced by using a computer and I will describe the use of a desktop or microcomputer in this area. There are several programs available for scanning and analyzing two-dimensional gels and I will mention these briefly, in passing. These programs need extensive computer facilities and a full discussion of these systems is beyond the scope of this chapter (*see* ref. *1* and references therein). One-dimensional gel scans, usually individual tracks from a slab gel, are routinely used for estimating the quantities of specific proteins, or the amounts of radioactivity in specific proteins. I will describe an interactive integration program that provides very flexible procedures for determining the areas under peaks in the scan. A more automatic integration system can also be used and a number of such systems, designed for chromatography, are available. I will describe a peak picking routine that could be used to develop an automatic integration program if commercial software is not available. One characteristic of gel electrophoresis is that the conditions of electrophoresis, unlike those in most chromatographic procedures, are close to equilibrium, so that the resultant band shapes are close to Gaussian. This means that a complex of overlapping bands may be resolved into its underlying Gaussian components. I discuss approaches to this problem and describe a flexible, robust program suitable for a good microcomputer. Finally, the comparison of gel scans involves transformations to correct for variations in run conditions and scan parameters. I describe a program that aligns and "stretches" gel scans and allows mathematical operations, such as generating the difference between two scans as a "difference scan."

Gel scans must be stored in a computer before they can be used. In principle, a chart record can be digitized, but this is slow and inaccurate. It is much better to record the scan directly into the computer, or onto a data-logger that gives a machine-readable output. Similarly, the programs have to be fed into the computer. The programs I will describe have been implemented in BASIC on Hewlett-Packard desktop computers. The BASIC listings are available from me or from the Hewlett-Packard Desktop Software Catalog.

Main Menu

The following options are available:

1. Scan a gel and store the scan.
2. Plot a stored scan.
3. Determine areas under peaks in a scan.
4. Resolve overlapping peaks into Gaussian components.
5. Compare two scans.

Options

Scan and Store

A digitized gel scan is a list of absorbance readings taken at equispaced positions, Dx, along the gel starting at a particular point, Xmin. The list of absorbance readings is stored together with the parameters Dx and Xmin and the number of readings in the list, Npts.

In some cases it is very useful to be able to digitally reduce the "noise" in a gel scan. This is particularly appropriate for very weak bands or where there is a high background, as in fluorography. Many methods to smooth curves are available, but the simpler ones, like a moving average, are not very effective and also distort the data. The best choice is a low pass digital filter that removes all the components that appear to be narrower than a set value, such as the grains on a photographic emulsion. Figure 1 shows an example of the use of a 0.1 dB Chebychev

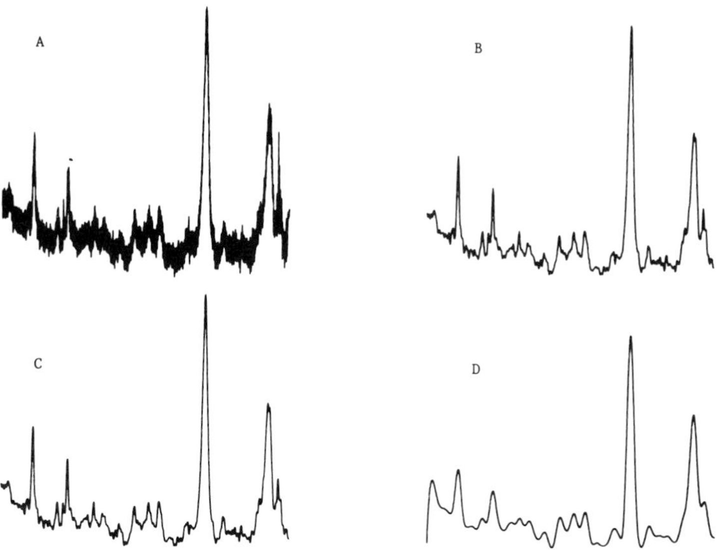

Fig. 1. Digital filtering. Scan A is the original data from a scan of a fluorograph of soluble histones (7). Scan B shows the result of passing the data through the digital filter with a relative cut-off frequency of 0.1. Scan C shows the result of reducing the number of data points in the filtered scan from 2840 to 568. Scan D shows the results of filtering the data of Scan A with a relative cut-off frequency of 0.02.

filter on a noisy gel scan. The program uses a second-order recursive filter, which is very stable, and passes the data through it four times to give the effect of an 8th-order filter. This technique is required for desktop computers because of their modest number precision. Phase shift is eliminated by passing the data alternately backwards and forwards through the filter.

After filtering, it may be possible to reduce the number of data points without affecting the quality of the scan. At this stage (or before filtering), it may be necessary to invert the scan on either axis. For example, a photographic negative of a stained gel shows the bands as peaks below a high background. This may be inverted by the statement MAT Scan = (Ymax) − Scan in Hewlett-Packard BASIC, where Scan is an array containing the absorbance values and Ymax is the maximum value in Scan. Inverting the *X*-axis may be necessary if some gels were scanned in different directions. Usually, gel scans are plotted with the top of the gel on the left. Inverting the *X*-axis requires a loop to move all the data points in the array as:

```
 FOR I  = 1 TO INT(Npts/2)      ! Lower subscript of
     J  = Npts − I + 1                  Scan is 1
  Temp  = Scan (I)
Scan(I) = Scan(J)
Scan(J) = Temp
 NEXT I
```

where I, J, and Temp are temporary variables and Npts and Scan have been defined above. Also at this stage it may be necessary to discard data points that are not needed at either end of the scan. This can be carried out by entering the *X*-axis values at the ends of the required part of the scan from the keyboard, or by digitizing on a plot of the scan. Digitizing in this application is achieved by placing a "+" or cursor on the scan as it appears on the graphics display. Keyboard controls are used to move the cursor, which moves along the scan, and hence to select the required section of the scan. The digitization subprogram is widely used in this set of programs to enter instructions. A modified version allows the cursor to be moved anywhere on the screen, in two dimensions, rather than being constrained to move along the scan.

Finally, changes to the baseline may be made, if desired.

The processed scan is then stored on a disk file and is available for analysis by subsequent programs. An integer format is used in order to conserve storage space. Sufficient descriptive information should be stored with the scan to allow it to be readily identified later.

Plot

A versatile plotting program is needed to prepare plots of the scans for discussion and for publication (e.g., 2). The program should allow for both rapid plotting, with minimal operator input, and for highly formatted plotting, where the user exercises extensive control over the plot. This is achieved by offering default parameters and making some sections of the program optional. Ideally, the plot should be developed on the CRT screen and then dumped, when perfect, to a graphics printer or a pen plotter. Figure 2 shows an example of a highly formatted plot, where the user controlled almost all of the parameters. Figure 1 is an example of plots produced automatically, with no user input. Axes may also be generated automat-

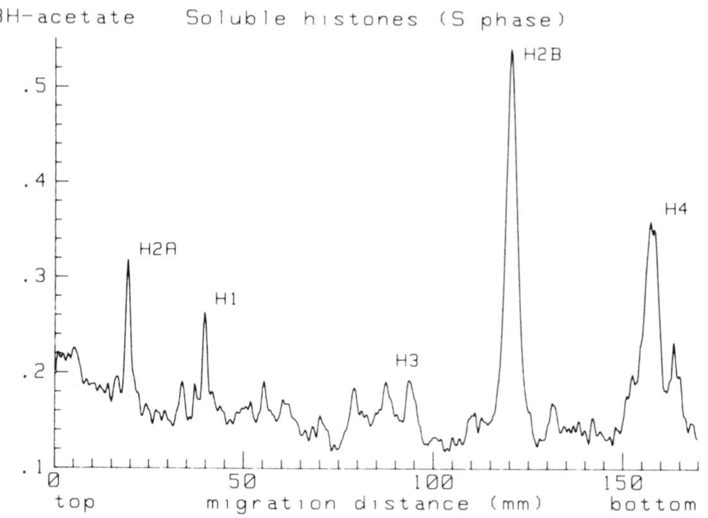

Fig. 2. Figure 1B replotted with axes and labels.

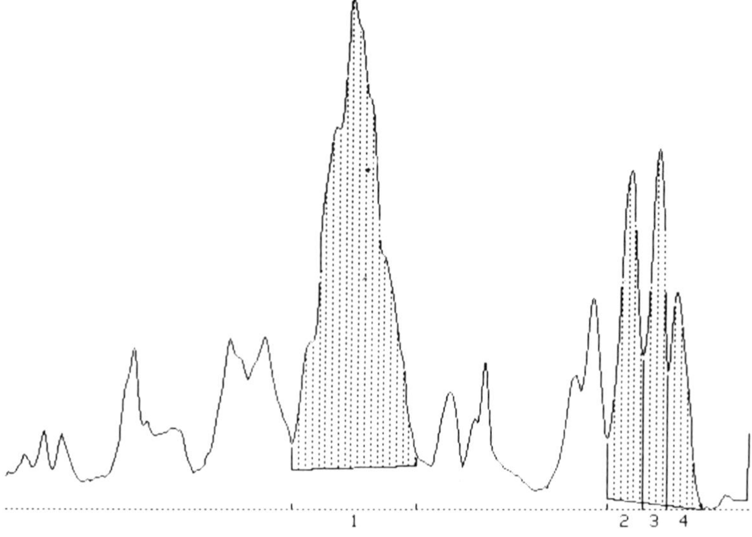

The range of this plot is .24mm to 172.44mm.

Demonstration of results of integration program.

Peak #	Area mm.abs.	Start mm.	End mm.	% of integrated area	normalized to band 1	normalized to total
1	25.4200	66.84	95.64	62.1	100	.62063
2	6.3751	140.04	148.14	15.6	25.08	.15565
3	5.4877	148.14	153.84	13.4	21.588	.13398
4	3.6755	153.84	161.94	9.0	14.459	.089738

Notes:
The last two columns show the areas normalized to either the area of the first integrated band (set to 100 from the keyboard) or to a total area of 1 (also entered from the keyboard).

Fig. 3. Integration of selected peaks on a gel scan. The integrated areas are shown by shading and the results are given in the table below the plot.

ically, if required. Figure 3 shows an example of a plot dumped to a graphics printer.

Areas

The main problem with determining areas on a gel scan is usually fixing the baseline. We use one of four options. Option 1 is to use the *X*-axis, which is arbitrarily drawn through the lowest point on the scan. Option 2 is to set a baseline by digitizing two points on the baseline and using them to define the baseline for the whole scan or part of the scan. Option 3 is to set the baseline as in option 2, but to set it separately for each area determined. Option 4 sets the baseline automatically for each area by drawing a straight line through the initial and final points of the area. Other programs for automatic integration may set the baseline by sensing parts of the scan where the first derivative is essentially zero.

A second problem with identifying the peaks interactively, i.e., with the digitizer is the lack of resolution of most microcomputer graphics displays. This is overcome by allowing the operator to choose part of the *X*-axis for display and automatically scaling the display so that the largest peak visible goes to the top of the screen. If only a small number of data points (< 100) is on the screen, they are joined by a cubic spline curve. The actual points are shown if there are < 50 points.

Each area determined is stored in memory with its associated parameters: starting *X*-value, finishing *X*-value, and baseline. The results are tabulated and related to standards by three methods: (i) each area is given as a percent of the total area determined; (ii) these percent values are multiplied by a constant; (iii) each area is given as a fraction of the area of a specified standard peak and these fractions may be multiplied by a constant. The table may be stored on disk and recalled for later study or additions or deletions.

The integration routine uses a cubic spline approximation to the peak if there are < 100 data points in the area chosen. For more than 100 points, trapezoidal integration is used in order to reduce memory requirements and increase speed. If the starting and finishing points lie between data points, then the starting and finishing points are determined by interpolation (usually cubic spline interpolation).

Overlapping Bands

There are general procedures available for resolving overlapping bands (e.g., *3*) and these must be used if non-Gaussian band shapes are encountered. However, in most cases the bands on electrophoresis gels that have not been overloaded or run too fast are Gaussian. Overlapping bands can be resolved by trial and error using a DuPont curve analyzer, which is an analog device that generates a composite of Gaussian curves that the operator compares with the gel band and then modifies the parameters of the Gaussian curves to get a good visual fit. A similar approach can be used on a digital computer (e.g., *4*). A digital computer may also be used to determine the parameters that automatically give a statistical best fit. The general approach is to minimize the sum of squares of differences between the actual absorbance values and the predicted absorbance values

$$\sum_{I=1}^{Npts} (\mathrm{Scan(I)} - \mathrm{Sum(I)})^2$$

where Sum is an array of predicted absorbance values for each of the Gaussians involved. A sum of Gaussians is mathematically nonlinear, so that an iterative process is used to minimize this function. A general nonlinear least squares fitting program can be used on a large computer (*5*) but, in my experience, the lower numerical precision found on microcomputers leads to problems in matrix inversion, so that the program frequently stops prematurely because of indeterminate equations. A simple intuitive procedure that works in simple cases is to fit one Gaussian, by linear least squares, in a region where it is not much affected by the overlapping Gaussians. Then extrapolate this fit into the overlap area and subtract it from the experimental scan. Then fit a second Guassian to the remainder. This type of approach has been used successfully in simple cases (*1*), but I have found it to be unstable when extended to several Gaussians and substantial amounts of overlap. I have used a general pattern-searching minimization program, which avoids the problem of matrix inversion instability, but it converged very slowly when run on the microcomputer used for this application. These difficulties led to the development of another minimization routine that has been found extremely stable and converges at least as rapidly as the less stable routines mentioned above. The routine minimizes the function, one parameter at a time, by calculating the function at the current value of the parameter and at two values a specified distance on either side of the current value. A parabola is drawn through the three points and the minimum value determined. This is repeated until the function is below a pre-set threshold or a given number of iterations is exceeded. The pattern of iteration is shown in the flow diagram (Fig. 4). This program makes it easy to choose some parameters from the keyboard and to introduce constraints such as making the bands equispaced and/or all the same width. It can also be advantageous to mix the automatic minimization with an interactive approach. For example, an operator can make changes in two parameters simultaneously and bypass quite a number of iterations of the automatic routine.

The initial input to this program is an estimate of the parameters of the Gaussians and the background. The minimization routine is not very sensitive to these and they can usually be determined by a peak-picking routine that runs through the first and second derivatives of the experimental scan to locate the peaks (first derivative zero, second derivative negative) and their widths (distance between points of inflexion, i.e., second derivative zero). The program allows the user to add, delete, or change Gaussians found by this peak-picking routine before entering the least squares minimization routine.

Output from the program is a table of the parameters of the Gaussians and their areas, and the parameters of the baseline plus plots of the experimental scan, the sum of Gaussians, the individual Gaussians, and the baseline. Figure 5 shows an example.

Compare

This program takes one scan as a standard scan, compares a second scan with it and, if required, generates a difference scan. First, the digitizer routine is used to identify a peak present on each scan that should be aligned. The computer aligns them. Second, the digitizer is used to indicate a second peak present on each scan that should be aligned. The computer scales (stretches or compresses the *X*-axis) one scan so that both pairs of peaks are aligned.

1. Check input parameters (avoid negative or zero width)

2. Initialize number of iterations

3. Increment number of iterations

4. Is the number of iterations >5? --- yes --→ 12.

5. Initialize number of new starts
this iteration

6. Calculate mean square error for

   (i) parameter value passed in
   (ii) parameter value passed in + increment
   (iii) parameter value passed in - increment

7. Print the parameter values and errors on the CRT monitor

8. Considering an imaginary parabola through the 3 points
(error, parameter value):

   (i) is there a maximum within the range? --- yes --→ 13.
   (ii) are the 3 calculated errors equal? --- yes --→ 13.
   (iii) is the minimum of the parabola outside the range? --- yes --→ 13.

9. There is a minimum in range so
Halve the increment for determining range & new parameter values &
Up-date the current best parameters

10. Is the current best error below threshold? --- yes --→ 12.

11. Was there a worthwhile reduction in error? --- yes --→ 3.

12. Print the best estimate
Pass out the new parameters

13. Increment the # of new starts

14. Make the best end of the old range into the center of the new range.

15. Calculate parameter for new range

16. Is # of new starts >2? --- yes --→ 12.

17. Calculate new mean square error

18. GOTO 7.

Fig. 4. Flow diagram for minimize routine.

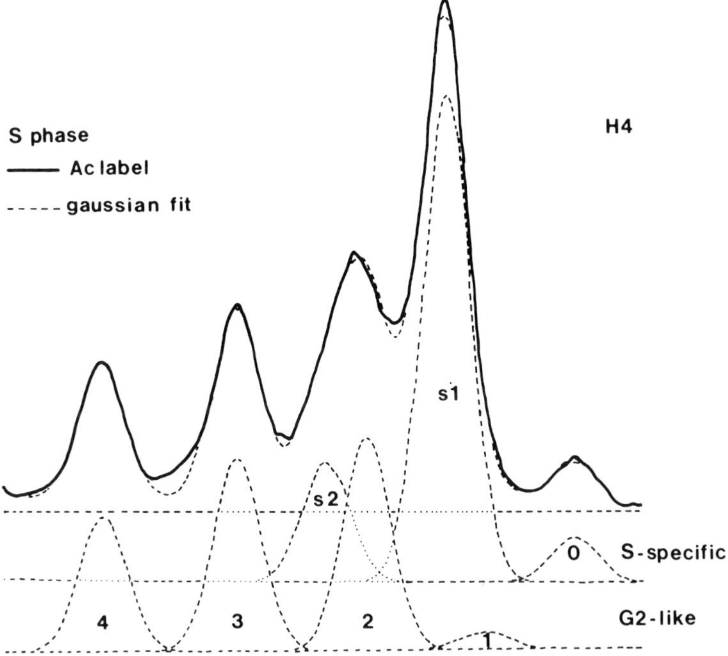

Fig. 5. Example of resolving a complex band into Gaussian components. The solid line shows the original scan and the broken line superimposed upon it shows the sum of the seven Gaussian bands shown below.

```
This program loads  an upper and a lower scan from disk.  The scans are
plotted and you can then align the lower scan with the upper scan.  The lower
scan can be subtracted from the upper scan and the subtracted curve becomes the
upper scan.

Program codes:     (Return to main program is 15)

Upper scan    load from disk (1)    plot (2)    store on disk (3)
Lower scan    load from disk (4)    plot (5)    store on disk (6)

Select a part of the scans (7)

Change plotting        offset (8)    scale(9)    line types (10)

X-axis        align (11)            scale (12)
Y-axis        scale (13)

Subtract lower from upper - difference curve replaces upper scan (14)

Enter program code required, then press ENTER.
```

Fig. 6. Menu for the program that aligns scans for comparison and calculates a difference scan.

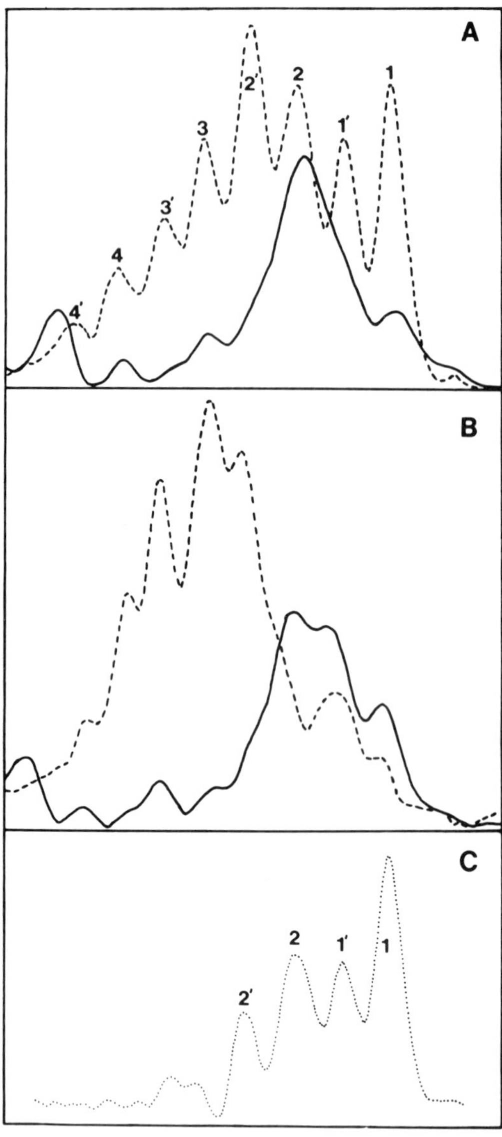

Fig. 7. Example of a difference scan. A. Part of a gel scan showing histone H3 pulse-labeled with ^{3}H in S phase of the cell cycle and separated on a Triton-acid-urea gel. Solid line: Coomassie blue stain. Broken line: fluorograph. B. As A except that the pulse of ^{3}H-acetate was given in G_2 phase of the cell cycle. C. Difference between the fluorograph scans in A and B. The numbers, 1, 1′, etc., refer to the numbers of acetate groups per H3 molecule in the band indicated.

Third, the digitizer is used to indicate a peak that should be the same height on each scan. The computer scales the *Y*-axis of one scan. Each of these processes is optional. To generate a difference scan, the program takes each data point on the lower scan, then interpolates on the upper scan to find the absorbance value at the same *X*-axis value and then subtracts the absorbance values. The resultant array is the difference scan.

The mathematical manipulation of gel scans can be extended. For example, Cooney et al. (6) corrected a pair of gel scans of DNA restriction fragments for impurities in the samples by solving a pair of simultaneous equations at each point along the gel.

Figure 6 shows the main menu for this program and Fig. 7 shows an example of a difference scan.

References

1. Garrison, J. C., and Johnson, M. L. (1982) A simplified method for computer analysis of autoradiograms from two-dimensional gels. *J. Biol. Chem.* **257,** 13,144–13,149.
2. Adler, M. (1982) A generalized plotting program, written in BASIC for scientific data. *Computer Programs Biomed.* **15,** 133–140.
3. Thomas, H. (1982) A procedure for resolving overlapping curves suitable for use with a microcomputer. *Anal. Biochem.* **120,** 101–105.
4. Chahal, S. S., Matthews, H. R., and Bradbury, E. M. (1980) Acetylation of histone H4 and its role in chromatin structure and function. *Nature* **287,** 76–79.
5. Murphy, R. F., Pearson, W. R., and Bonner, J. (1979) Computer programs for analysis of nucleic acid hybridization, thermal denaturation, and gel electrophoresis data. *Nucleic Acids Res.* **6,** 3911–3921.
6. Cooney, C. A., Matthews, H. R., and Bradbury, E. M. (1984) 5-Methyldeoxy-cytidine in the *Physarum* minichromosome containing the ribosomal genes. *Nucleic Acids Res.,* **12,** 1501–1515.
7. Waterborg, J. H., and Matthews, H. R. (1984) Patterns of histone acetylation in *Physarum polycephalum*: H2A and H2B acetylation is functionally distinct from H3 and H4 acetylation. *Eur. J. Biochem.* **142,** 329–335.

Chapter 16

Drying Gels

Bryan J. Smith

Institute of Cancer Research, Royal Cancer Hospital, Chester Beatty Laboratories, London, England

Introduction

There are several reasons why it may be desirable to dry a gel after its use for electrophoresis. Firstly, it is a convenient way of storing the end result of the experiment. Secondly, drying a gel that is fragile may make it easier to handle (say, during optical scanning). Thirdly, and most importantly, it may be necessary to dry a gel to allow the most efficient detection of radioactive samples on it by autoradiography or fluorography. The method described below is rapid and generally applicable to all types of slab gel, although the description below applies specifically to a thin (0.5–1 mm thick) polyacrylamide (15%) gel being prepared for fluorography (*see* Chapter 17).

Materials

1. A high vacuum, oil pump. Protect the pump and its oil from acid and water by inclusion of a cold trap in the vacuum line.

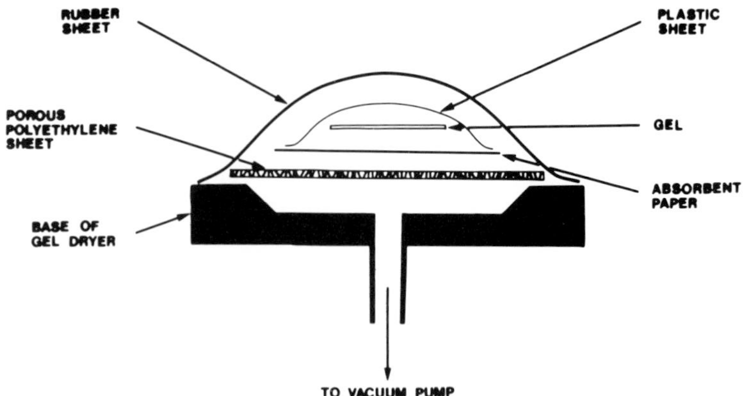

Fig. 1. Diagram showing the construction of a simple gel dryer.

2. A heat source (such as a hot air fan, infrared lamp, electric hot plate, or a steam/hot water bath).
3. A gel dryer, available commercially from various sources or readily made in the laboratory, along the lines of the dryer shown diagrammatically in Fig. 1. Essentially, a gel dryer has the gel placed on top of absorbent paper, which in turn is supported by a firm sheet of porous polyethylene and/or a metal grille. As shown in Fig. 1, the gel is covered with "plastic sheet," such as Saran Wrap or Cling Film transparent domestic food wrapping. This construction is put under vacuum beneath a sheet of silicon rubber and is heated so as to help drive off moisture. Suitable polyethylene sheet, absorbent paper, cellophane, and other items may be obtained commercially. Preferably the absorbent paper, to which the dried gel adheres, should be about 1 mm thick. If it is thinner (as is, say, Whatman 3 MM paper), it has a tendency to curl up once removed from the gel dryer.

Method

1. The gel will already have been run and, if carrying out fluorography, suitably treated and then washed thoroughly with distilled water (*see* Chapter 17).

2. Place the gel on a piece of absorbent paper that is slightly larger than the gel itself. Do not trap air bubbles between them. Place them, gel uppermost, on top of the porous polyethylene sheet and cover the whole with some of the nonporous Cling Film plastic film. Place this construction on the gel dryer base and cover over with the sheet of silicon rubber, as shown in Fig. 1.
3. Apply the vacuum, which should draw all layers tightly together. Check that there is no air leak. After about 10–15 min, apply heat (say, 60°C) evenly over the face of the gel.
4. Continue until the gel is dry, at which point the silicon rubber sheet over the gel should assume a completely flat appearance. The time taken to dry a gel is dependent on various factors (see Notes), but a gel of 0.5–1 mm thickness, equilibrated with water, will take 1–2 h to dry.
5. When the gel is completely dry and bound onto the absorbent paper, it may be removed and the Cling Film over it peeled off and discarded.

Notes

1. The plastic Cling Film layered over the gel is used for two reasons. Firstly, it prevents sticking of the gel to the rubber sheet covering it. Secondly and more importantly, it reduces the likelihood of contamination of the dryer by radioactive substances from the gel. After drying, the plastic film may be discarded as radioactive waste.
2. If a suitable high vacuum pump is not available for use with the gel dryer, a good water pump may suffice instead, but in this case the gel will take longer to dry. As another alternative, the gel may be dried in atmospheric conditions, for instance as described in (*1*). In that case the gel is equilibrated with 2% (v/v) glycerol in water, covered with porous cellophane, and kept flat while it dries in the air at room temperature. This process may take a day or so.
3. Gels for storage can be dried, as for fluorography, onto paper. However, gels for transmission optical

scanning obviously need to be transparent. For this purpose the gel is sandwiched between two sheets of porous cellophane (with no trapped air bubbles), and then taken through the process described above. The dry gel will not adhere to the absorbent paper.

4. The main problem with this method is that gels may crack up and so spoil the end result. This may occur as a chronic process, during the drying. Gels of higher %T acrylamide and of greater thickness are particularly prone to suffer this fate. However, to alleviate this problem the gel may be treated before drying with one of several solutions, which are:

   (i) 70% (v/v) methanol in water (*2*). Equilibrate the gel by soaking in this solution (0.5–1 h for thin, 1-mm, gels or longer for thick gels) and then proceed with drying. This treatment will cause the gel to shrink. High %T gels may dehydrate and go opaque. This process may be reversed with water, but if it is too extreme the gel may crack. If this is a danger, use a weaker methanol solution (say, 40% v/v). This treatment speeds up the drying process somewhat since the methanol is driven off fairly quickly.
   (ii) Glycerol (1% v/v) and acetic acid (10% v/v) in water (*3*). Equilibrate the gel and proceed with drying.
   (iii) DMSO (2% v/v) and acetic acid (10% v/v) in water (*3*). Equilibrate the gel and proceed with drying.

   Of these three, the DMSO solution is most likely to prevent cracking of difficult gels and the 70% methanol the least likely, but the former will take the longest time to dry down, and the latter the least time.

5. As a further precaution, and also to speed up drying, a second sheet of porous polyethylene may be used, so that the gel, *without* the overlaid, nonporous plastic Cling Film sheet, is sandwiched between the two polyethylene sheets. This arrangement provides a greater surface of gel for drying. However, ensure that the face of the polyethylene sheet that is in contact with the gel is very smooth, for otherwise the gel

will dry into it (as well as into the absorbent paper) and they will be difficult to separate. If this remains a problem, employ a sheet of porous cellophane between the gel and polyethylene. The cellophane may be removed after drying.

6. Another precaution is to use a lower temperature during the drying process, so that gradients of temperature and hydration through the system are less extreme.

   Thus, probably the best approach to drying a difficult gel such as a 3 mm-thick 10% or 15%T acrylamide gel would be to use the DMSO (2%) soaking of the gel before drying, and two polyethylene sheets and less heat (say, 40°C) during the drying. Under these circumstances, a thick gel may take a whole working day to dry down.

7. Gel cracking may also occur as an acute phenomenon when the vacuum is released from a gel that is not completely dry. Thus, it is important to check for air leaks in the dryer and not to end the drying (i.e., release the vacuum) too soon. So that this does not happen, it is important to determine (by trial) the time required to dry down a gel in one's own gel-drying equipment. This time increases with increased %T of acrylamide, thickness and surface area of the gel, and with decreased temperature and vacuum during the drying.

8. A problem may arise if the gel is prepared for fluorography using DMSO as solvent. The gel must be thoroughly washed in water (or other solution, see Note 4) to remove the DMSO. This is because DMSO has a boiling point of 189–193°C and is difficult to remove under the conditions of drying. If remaining in the gel in significant amounts, the gel will remain sticky and the photographic film it contacts may become fogged.

9. Agarose gels may be dried in the manner described. Even 3-mm agarose gels dry quickly (about 0.5 h) and without cracking when using only one polyethylene sheet and room temperature for drying. Composite weak acrylamide–agarose gels are likewise readily dried down. When dealing with gels containing

agarose, beware the use of DMSO, which dissolves agarose.

10. If using EN^{3}HANCE (New England Nuclear) in the gel, do not employ temperatures above 60°C for drying, for EN^{3}HANCE is volatile above 65°C.

References

1. Giulian, G. G., Moss, R. L., and Greaser, M. (1983) Improved methodology for analysis and quantitation of proteins on one-dimensional silver-stained slab gels. *Anal. Biochem.* **129,** 277–287.
2. Joshi, S., and Haenni, A. L. (1980) Fluorographic detection of nucleic acids labelled with weak β-emitters in gels containing high acrylamide concentrations. *FEBS Lett.* **118,** 43–46.
3. Bio-Rad, Bulletin 1079 (1981) Model 1125B high capacity gel slab dryer for protein gels and DNA sequencing.

Chapter 17

Fluorography of Polyacrylamide Gels Containing Tritium

Jaap H. Waterborg and Harry R. Matthews

Department of Biological Chemistry, University of California School of Medicine, Davis, California

Introduction

Fluorography is the term used for the process of determining radioactivity in gels and other media by a combination of fluorescence and photography. Since most of the radiation of a low energy emitter will largely be absorbed by the gel, in the technique of fluorography a fluor (e.g., PPO) is infiltrated into the gel where it can absorb the radiation and re-emit light that will pass through the gel to the film. The resulting photographic image is analogous to an autoradiograph, but for a low energy β-emitting isotope like ^3H, the sensitivity of fluorography

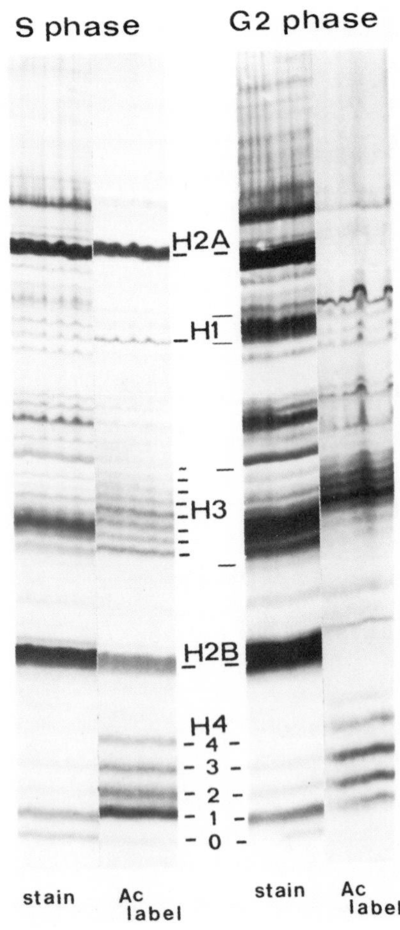

Fig. 1. This shows two examples of fluorography of protein bands labeled with ^{3}H. Basic nuclear proteins were isolated from the slime mold, *Physarum polycephalum*, pulse-labeled with ^{3}H-acetate in either S phase or G_2 phase of the naturally synchronous cell cycle. The proteins were analyzed by acrylamide gel electrophoresis in acetic acid, urea, and Triton X-100. After electrophoresis, the gel was stained with Coomassie blue, photographed, and then fluorographed. Individual lanes of the gel image were cut from the photograph (negative) and the fluorograph and then printed side-by-side to give the figure shown.

Notice that the stain patterns of the two lanes are practically identical, except for the loading, while the radioactivity patterns show major differences, for example, the absence of label in histones H2A and H2B in G2 phase (4).

is many times the sensitivity of autoradiography. The fluorograph may be used directly, as a qualitative picture of the radioactivity on the gel. It may also be used to locate radioactive bands or spots that can then be cut from the original gel for further analysis, or be scanned to give quantitative information about the distribution of radioactivity. Figure 1 shows an example of a gel that was stained with Coomassie blue and then fluorographed. Notice that there is no loss of resolution in the fluorography of thin gels of normal size.

The procedures described here are based on those described by Laskey and Mills (*1*), Bonner and Laskey (*2*), and Randerath (*3*).

Materials

1. −70°C freezer
2. Film cassette, preferably with enhancing screen
3. Small photographic flash unit
4. Gel dryer
5. Film developer
6. Fixing solution: 7% acetic acid, 20% methanol in distilled water.
7. Acetic acid, 25% (v/v)
8. Acetic acid, 50% (v/v)
9. Acetic acid (glacial), 100%
10. PPO solution: 20% (w/v) PPO (2,5-diphenyloxazole) in glacial acetic acid.
11. Film: Kodak XAR-5

Method

1. After electrophoresis, remove the gel from the apparatus and fix by soaking in Fixing solution for 1 h. If required, the gel may then be stained with Coomassie blue, destained, and photographed.
2. Dehydrate the gel by shaking it for 10 min each in 25% acetic acid, 50% acetic acid, and glacial acetic acid.

3. Completely cover the gel with not less than 4 gel volumes of PPO solution and shake the gel in the PPO solution for 2 h.
4. Transfer the gel gently and evenly to a dish of distilled water and shake for 2 h. In water, PPO precipitates so that the gel turns opaque white.
5. Dry the gel completely (*see* Chapter 16), using Saran wrap on one side, on Whatman 3MM paper.
6. Open a film cassette and place the dry gel (after removal of the Saran wrap) in the cassette with the enhancing screen if used. Leave the top of the cassette beside the part with the gel in it with the white inner lining facing up.
7. In complete darkness, place a sheet of film on the white lining of the open cassette and pre-flash it. (*see* Note 5)
8. Assemble the cassette with the film directly on the gel. Wrap the cassette with aluminum foil and place in the −70°C freezer.
9. Remove the cassette from the freezer after the appropriate exposure time. (*see* Note 7) and allow to warm up for about 2 h at room temperature.
10. In complete darkness, open the cassette and develop the film.

Notes

1. Other fixing solutions may be used in step 1. For example, formalin can be used to fix peptides covalently in the gel.
2. Coomassie blue staining gives minimal color quenching, but very heavily stained bands may show reduced efficiency for fluorography. Amido black gives more color quenching and is not recommended. Silver staining has not been tested. It is possible to re-swell the gel after fluorography and stain it then, but some loss of resolution occurs.
3. The times given in steps 2 and 3 are for 0.5–1.5 mm thick gels containing 15% acrylamide. Thicker or more concentrated gels will require longer periods in all solutions.

4. Since PPO is expensive, it is normally recycled as follows: Set up four 4 L Erlenmeyer flasks with a large stirring bar and about 2.5 L of distilled water in each. Add about 0.25 L of used PPO solution, slowly, to each Erlenmeyer, stirring continuously. The PPO crystallizes out. Collect the PPO by filtering the solutions. Dry the crystals at 20°C for 2–3 d. Dissolve the crystals in a minimum volume of ethanol and precipitate, filter, and dry again. Finally, dry the PPO in a vacuum oven for about 1 wk, breaking up lumps at intervals. Glassware that was used for PPO should be rinsed in ethanol before washing.
5. *Pre-flashing* is used to improve the sensitivity and linearity of response of the film (*1*). Use a small, battery-operated, photographic flash unit. Tape a red filter and a diffusing screen of Whatman 3 MM paper over the flash window. Experiment with the number of layers of paper required to give an absorbance of about 0.1 when the film is pre-flashed. To pre-flash, hold the flash unit directly above the film, about 60 cm away, and press the manual flash button. Use a procedure that you can easily reproduce in the dark. Note that the first flash after the flash unit is switched on may be different from subsequent flashes, so avoid using the first flash. The white inner lining of the film cassette provides a uniform, reproducible background for pre-flashing.
6. Be careful not to place the cassette near penetrating radiation from sources such as a ^{32}P or ^{125}I autoradiograph or radioactive samples. This radiation will fog the fluorograph. Make sure the cassette is light-tight, too.
7. Recommended exposure times are of the order of 24 h for 1000–10,000 dpm ^{3}H and correspondingly longer for lower amounts of radioactivity. Exposure times of several months do not give a significant increase in background. Sensitivity for ^{14}C is reported to be about 10× that for ^{3}H. Note that high acrylamide concentration such as the 50% gels used for peptide analysis severely quench the fluorescence and drastically reduce the efficiency of fluorography.
8. An automatic developer is most convenient. Manual development is described in Vol. 2, Chapter 8.

References

1. Laskey, R. A., and Mills, A. D. (1975) Quantitative film detection of ^{3}H and ^{14}C in polyacrylamide gels by fluorography. *Eur. J. Biochem.* **56,** 335–341.
2. Bonner, W. M., and Laskey, R. A. (1974) A film detection method for tritium-labelled proteins and nucleic acids in polyacrylamide gels. *Eur. J. Biochem.* **46,** 83–88.
3. Randerath, K. (1970) An evaluation of film detection methods for weak β-emitters, particularly tritium. *Anal. Biochem.* 34, 188–205.
4. Waterborg, J. H. and Matthews, H. R. (1983) Patterns of histone acetylation in the cell cycle of *Physarum polycephalum*. *Biochemistry* **22,** 1489–1496.

Chapter 18

Recovery of Proteins from Dried Polyacrylamide Gels after Fluorography

Jaap H. Waterborg and Harry R. Matthews

Department of Biological Chemistry, University of California School of Medicine, Davis, California

Introduction

Several methods are available for recovering proteins electrophoretically from polyacrylamide gels directly after electrophoresis or after staining for protein (*see*, for example, Chapter 19). Fluorography (*see* Chapter 17) is used to detect small amounts of proteins or specifically modified forms of proteins by labeling these with specific precursors such as ^{3}H-lysine or ^{3}H-acetate.

However, fluorography requires dried gels and thus prevents the use of existing methods for protein recovery from gels. We have found that dried gels containing PPO

(2,5-diphenyloxazole) can rapidly be reswollen in acetic acid. The gels may then be eluted electrophoretically so that the labeled (and also the unlabeled) proteins can be analyzed for amino acid composition, tryptic peptide analysis, and even Edman degradation to determine amino acid sequence (*1*,*2*)

Materials

1. Swelling solution: glacial acetic acid with 0.001 % (w/v) Coomassie Brilliant Blue.
2. Equilibration buffer (*see* Chapter 19): 1*M* acetic acid, 50 m*M* NaOH, 1% cysteamine.

Method

1. Determine the position of the protein band(s) of interest in the polyacrylamide gel, dried on Whatman 3MM paper, after fluorography (Chapter 17).
2. Use scissors or a razor blade to cut out the required piece of dried gel together with its paper backing.
3. Place up to 10 mL of gel pieces (*see* Note 4) in a 50 mL polypropylene tube with screw cap. Add 10 volumes of swelling solution to the tube, and gently rock, roll, or invert for 15 min.
4. Decant the solution from the swelling gel pieces and discard.
5. Repeat steps 3 and 4 once.
6. Equilibrate the swollen gel pieces for 30 min in 10 volumes of equilibration buffer, decant the equilibration buffer from the gel pieces, and discard.
7. Repeat step 6 once and use the reswollen and equilibrated gel pieces in the electrophoretic elution procedure described in Chapter 19.

Notes

1. This protocol for reswelling dried and fluorographed polyacrylamide gels is ineffective on dried gels that do not contain PPO.

2. Labeled protein bands can be precisely located in the dried gel by placing the fluorograph film on top of the dried gel. Generally, a marker pen injected with some ^{14}C-labeled compound is used to mark the filter paper at several positions next to the gel with sufficient cpm to give a small dark spot on the film after the fluorography. In gels previously stained with Coomassie Brilliant Blue, the residual stain can assist in precisely superimposing the film on the gel. Although gel staining prior to fluorography is not required when only labeled proteins are to be recovered, unlabeled proteins can only be isolated when they can be localized by their residual stain.
3. Cut the pieces of gel-on-paper small enough to assure homogeneous swelling, e.g., diameter less than 0.5 cm.
4. The "volume" of the pieces is measured as a loose layer. During the first wash the gel pieces will be seen to rapidly swell to up to twice the original gel thickness.
5. The gel pieces, in equilibration buffer, will generally display a white center of reprecipitated PPO. Gel pieces from thick gels may even be completely white. This PPO will not interfere with the subsequent electrophoretic elution, and reproducible and complete protein recovery is obtained when the gels are completely reswollen. For very thick gels (more than 2 mm thick) steps 4 and 5 may need to be repeated twice, and care should be taken to reduce the size of the gel pieces as much as possible. Loss of protein may occur wen the number of washes in swelling solution is increased, especially when the trace of Coomassie is omitted from the swelling solution.

References

1. Mende, L. M., Waterborg, J. H., Mueller, R. D., and Matthews, H. R. (1983) Isolation, identification and characterization of the histones from plasmodia of the true slime mold, *Physarum polycephalum,* using extraction with guanidinium hydrochloride. *Biochem.* **22,** 38–51.
2. Waterborg, J. H., and Matthews, H. R. (1983) Acetylation sites in histone H3 from *Physarum polycephalum. FEBS Lett.,* **162,** 416–419.

Chapter 19

The Electrophoretic Elution of Proteins from Polyacrylamide Gels

Jaap H. Waterborg and Harry R. Matthews

Department of Biological Chemistry, University of California School of Medicine, Davis, California

Introduction

The analytical power of acrylamide gel electrophoresis is one of the keys of modern protein chemistry. It is not surprising, therefore, that many methods have been described for converting that analytical power into a preparative tool. None of the available methods is entirely satisfactory for general use since loss of resolution or low recovery is often involved. The method described here has given both high resolution and good recovery but suffers from the disadvantage of being relatively laborious (*1,2*). In addition, although the recovered proteins are good for peptide analysis or amino acid composition determination, we have found very low yields on Edman degradation of proteins eluted from gels (*3*).

The method described below works well for eluting proteins from acid–urea–Triton gels and should work equally well with acid–urea gels (Chapter 8). Wu et al. (*1*) describe an alternative set of buffers that can be used for SDS gels.

The method uses the principle of isotachophoresis, as in the stacking gel portion of a discontinuous gel electrophoresis system (*see*, for example, Chapter 6). The gel pieces containing the protein of interest are embedded in agarose above an agarose gel column. A detergent, CTAB, is used to displace the Triton (or SDS) bound to the protein that is electrophoretically eluted into the agarose gel column. In the column, the protein is concentrated by stacking between the leading ion (Na^+) and the trailing ion (betaine). The protein dye (Coomassie blue or Amido black) stacks ahead of the protein. The concentrated protein band is cut from the agarose gel and recovered.

Materials

1. Lyophilizer
2. Vacuum pump
3. Electrophoresis apparatus capable of running tube gels, preferably in glass tubes with small funnels like the BioRad Econocolumn #737-0243 (which is 5 mm inner diameter and 20 cm long) with the lower end cut off.
4. Staining solution (Coomassie): 0.1% (w/v) Coomassie Brilliant blue R, 5% (v/v) acetic acid, 40% (v/v) ethanol, 0.1%; (w/v) cysteamine in distilled water.
5. Destaining solution (Coomassie): 5% (v/v) acetic acid, 40% (v/v) ethanol, 0.1% (w/v) cysteamine.
6. Staining solution (Amido): 0.1% (w/v) Amido black in destaining solution (Amido) (note: add the cysteamine immediately before use).
7. Destaining solution (Amido): 7% (v/v) acetic acid, 20% (v/v) methanol, 0.1% (w/v) cysteamine.
8. Equilibration buffer: 1*M* acetic acid, 50 m*M* NaOH, 1% (w/v) cysteamine.

9. Siliconizing solution: 1% (v/v) Prosil-28 in distilled water.
10. LMT agarose solution: 0.5% (w/v) low melting temperature agarose in 1*M* acetic acid, 50 m*M* NaOH.
11. HMT agarose solution: 1% (w/v) high melting temperature agarose, 1*M* acetic acid, 50 m*M* NaOH, 0.0005% (w/v) methyl green in distilled water (*see* note 3).
12. Upper reservoir buffer: 1*M* acetic acid, 0.1*M* betaine, 0.15% (w/v) Cetyl Trimethyl Ammonium Bromide (CTAB) in distilled water. Lower reservoir buffer: 1*M* acetic acid, 50 m*M* NaOH.
13. Acidified acetone: add conc. HCl to acetone to give a concentration equivalent to 0.02*N*.
14. Elution buffer: 0.02*N* HCl with optionally 0.1% (w/v) cysteamine, 0.1*M* *N*-methyl-morpholine-acetate, pH 8.0
15. Agarose gels: (can be done 1 d before needed):
    (a) Thoroughly clean and dry the tubes and then siliconize them, for example, by immersing in 1% Prosil-28 followed by thorough rinsing with water and drying.
    (b) Soak small pieces (3 cm square) of dialysis membrane in distilled water and fix one piece over the end of each tube with an elastic band. This is intended to hold the agarose in the tube.
    (c) Melt LMT agarose solution by heating to 65–100°C. Use 7 mL per tube. Fill each tube carefully, avoiding trapping air bubbles, just to the bottom of the funnel. Let the agarose gel at 4°C.

Method

1. Stain the gel just enough to visualize the band(s) of interest. Either Coomassie blue or Amido black staining may be used. If the gel is loaded heavily enough, then 15 min in either staining solution and no destaining is recommended.
2. Cut out the bands of interest and soak them for at least 1 h, or overnight, in equilibration buffer.
3. Chop the gel bands into small pieces with a razor blade and transfer them to the funnels on the tops of the agarose gels in the electrophoresis apparatus.

4. Melt the HMT agarose solution in a 100°C bath and add between 2 and 5 mL to each funnel. Stir to remove air bubbles and achieve a uniform distribution of gel pieces in the agarose.
5. Let the agarose solution gel either at room temperature or at 4°C.
6. Place the bottom reservoir of the electrophoresis apparatus in a tray containing melting ice as coolant. Fill the upper and lower reservoir with upper reservoir buffer and lower reservoir buffer, respectively. Connect the negative terminal of the power supply to the lower electrode and the positive terminal to the upper electrode.
7. Start the electrophoresis, using 2.5–5 mA per tube (about 200 V).
8. During electrophoresis, the staining dye (Coomassie or Amido) migrates fastest, followed by the buffer discontinuity, the protein, and the methyl green marker dye. Continue electrophoresis at least until the staining dye front reaches the middle of the gel. If the buffer discontinuity is clearly separated from the staining dye, then proceed to step 9. Otherwise, continue electrophoresis until the staining dye is clearly separated.
9. Immediately remove the gel from the tube as follows:
   (a) Use a spatula to remove the acrylamide gel pieces and their supporting HMT agarose gel.
   (b) Push the gel column with a plunger from a disposable syringe out of the tube onto a clean glass plate.
10. Immediately cut the gel about 3 mm below the buffer discontinuity (above the staining dye) and about 5 mm above the buffer discontinuity. The resulting 8 mm section of gel contains the protein and the methyl green marker dye; the remainder is discarded. Place the gel section in a 1.5 mL polypropylene centrifuge tube.
11. Use the following procedure (steps 12–15) if the presence of agarose, and protein denaturation during drying, do not interfere with subsequent analysis. Otherwise proceed to step 16.

12. Add 1 mL of acidified acetone and leave overnight in the freezer, at −20°C.
13. Centrifuge (10,000*g*; 5 min) and slowly decant the acetone, which contains the dye, buffer salts, and detergent.
14. Cap the tube, pierce the cap, and dry the gel under vacuum using a vacuum pump until the vacuum falls below 100 mtorr.
15. Rehydrate the agarose with 20–50 μL of distilled water, or appropriate buffer, by incubation at 65°C for a few minutes. This solution will remain liquid at 37°C, allowing enzymatic digestion, or it may be diluted to give less than 0.1% agarose when it will no longer gel. This is the end of this procedure.
16. This step follows step 11 if steps 12–15 were unsuitable. If this procedure is used, then the methyl green marker dye should not be used (*see* note 3).
17. To the gel slice, add 0.5 mL elution buffer. Mix very gently without breaking up the agarose. Leave at room temperature for at least 1 h and preferably overnight.
18. Equilibrate a small (~1 × 10 cm) desalting column of Sephadex G-25 with 0.1*M* *N*-methylmorpholine acetate, pH 8.0.
19. Centrifuge (10,000*g*; 5 min). Carefully decant or pipet off the supernatant and set it aside.
20. Repeat steps 17 and 19, combining the supernatants. The extracted gel may be discarded.
21. Load the combined supernatants onto the Sephadex desalting column equilibrated with 0.1*M* *N*-methylmorpholine acetate, pH 8.0 (step 18). Elute with 0.1*M* *N*-methylmorpholine acetate and collect the material eluting in the excluded volume.
22. Lyophilize the protein to dryness. The resulting protein is salt-free since *N*-methylmorpholine acetate is volatile. This ends the alternative procedure.

Notes

1. Gels that have been normally stained and destained may be used but the time of agarose gel electrophore-

sis may have to be increased or the capacity of the system will be reduced. (Note that cysteamine should not be included for Coomassie staining of histone H1.)

2. For Amido black, but not for Coomassie, the equilibration step (step 2) also acts as a destaining step.
3. The presence of methyl green in the HMT agarose makes it easy to locate the buffer discontinuity during the subsequent electrophoresis and is mostly removed by acetone precipitation of the protein. Small residual amounts do not interfere with subsequent peptide mapping. However, if the protein is to be isolated and desalted by the Sephadex method (step 16 on) the methyl green should be omitted from the HMT agarose solution.(If methyl green *were* included it would elute partly with the protein and partly immediately after the protein). In this case the buffer discontinuity must be located directly from the refractive index change.
4. At step 4, use the minimum volume of agarose required to suspend the gel fragments since the time of electrophoresis depends greatly on the volume of agarose plus gel fragments.
5. Betaine (rather than glycine) is used as the trailing ion because of its higher acetone solubility. It also increases the capacity of the system.
6. If ice cooling is not used during electrophoresis, reduce the current to 2.5 mA/tube.
7. The distance between the lower edge of the staining dye and the buffer discontinuity depends on the amount of stain and how full the funnels are on top of the gel, varying from about 5 to about 40 mm. The distance between the upper edge of the staining dye and the buffer discontinuity is fairly constant at 5–10 mm (reduces to 2–3 mm if the betaine buffer is replaced by glycine).
8. The time of electrophoresis will be 2–4 h at 4 mA/tube for low amounts in the funnels, and up to overnight at 2.5 mA/gel for full funnels. Different tubes may require different running times. In this case, remove the tubes whose electrophoresis is complete as follows: (a) turn off the power, disconnect the power supply, and empty the upper reservoir buffer into a beaker.

(b) Remove the required tube(s), block the resulting hole(s) with rubber bung(s), replace the upper reservoir buffer, and continue the electrophoresis, adjusting the power supply as necessary.

If large volumes of reservoir buffers (e.g., 2 L) are used, then reservoir buffer changes are not needed.

9. It is important to remove the gel immediately after eletrophoresis since the staining dye diffuses rapidly once the current is turned off. It may be possible to remove the gel in other ways (e.g., *see* ref. *1*), but we have found the method described in step 9 to be rapid and reliable.
10. Slice the gel immediately after electrophoresis, to prevent loss of resolution by diffusion. In our experience, this 8 mm section of the gel contains at least 98% of the protein.
11. The simple acetone extraction and drying procedure (steps 12–15) is suitable if the protein is to be digested with enzymes and the products analyzed by gel electrophoresis. Note that the procedure described by Cleveland (*see* Chapter 22) provides another peptide mapping approach for some applications. The alternative procedure (step 16 on) is recommended if the protein is to be analyzed for amino acid composition or characterized by nuclear magnetic resonance or other techniques, for example, thin layer analysis of tryptic digests.
12. The protein will precipitate in the gel at step 12, possibly forming a white band, but precipitation may not be quantitative with extremely small amounts of protein.
13. Drying of the extracted gel (step 14) takes several hours. This procedure may result in irreversible denaturation of the protein, but it will probably still be solubilized by enzymic digestion.
14. At step 17, do not vortex or mix vigorously or you will get agarose in the final protein solution. Allow the protein to diffuse out. Do not include urea or other agents that would solubilize the agarose in the elution buffer.
15. The second extraction (step 20) need only be 1 h. In our tests, less than 5% of the protein remained in the

gel after the two extractions, although the white appearance persisted. The protein concentration in the eluate cannot be determined by the Bradford method (*4*) since CTAB strongly interferes.

References

1. Wu, R. S., Stedman, J. D., West, M. H. P., Pantazis, P., and Bonner, W. M. (1982) Discontinuous agarose electrophoretic system for the recovery of stained proteins from polyacrylamide gels. *Anal. Biochem.* **124,** 264–271.
2. Mende, L. M., Waterborg, J. H., Mueller, R. D., and Matthews, H. R. (1983) Isolation, identification and characterization of the histones from plasmodia of the true slime mold, *Physarum polycephalum,* using extraction with guanidinium hydrochloride. *Biochemistry* **22,** 38–51.
3. Waterborg, J. H., and Matthews, H. R. (1983) Acetylation sites in histone H3 from *Physarum polycephalum. FEBS Lett.,* **162,** 416–419.
4. Bradford, M. M. (1976) A rapid and sensitive method for quantitation of microgram quantities of protein utilizing the principle of protein–drug binding. *Anal. Biochem.* **72,** 248–254.

Chapter 20

Transfer Techniques in Protein Blotting

Keith Gooderham

MRC Clinical and Population Cytogenetics Unit, Western General Hospital, Crewe Road, Edinburgh, United Kingdom

Introduction

Polyacrylamide gel electrophoresis is an extremely powerful tool for the analysis of complex protein mixtures. Although the value of this method cannot be questioned, it is restricted in that the separated proteins remain buried within the dense gel matrix and are not readily available for further investigation. A number of methods have been developed in order to try and overcome this problem, for example the elution of proteins from excised gel slices (*see* Chapter 19). Alternatively, proteins have been studied while they are still buried within the gel using a variety of *in situ* peptide mapping (*see* Chapter 22) and gel overlay techniques (for example, *see* ref. *1*). Unfortunately all of these methods have serious drawbacks: in the case of protein elution and *in situ* peptide mapping techniques, the resolution and number of bands that can be processed is restricted, whereas the gel overlay techniques are generally time-consuming and insensitive.

Recently a new technique—protein blotting—has been developed that promises to overcome many of these problems (*see* ref. *2* for a recent review). Using this method, proteins are transferred out of the gel and onto a filter or membrane, forming an exact replica of the original protein separation, but leaving the transferred proteins accessible to further study. A wide range of different probes, including antibodies, DNA, RNA, and lectins, have been used in protein blotting experiments and the potential applications of this method are only limited by the availability of suitable probes and assay systems.

As has already been emphasized, the great advantage of protein blotting is that the transferred proteins are no longer trapped within the gel, but are freely accessible to further analysis by being bound to the surface of a filter. It is clear therefore that the transfer step is the crucial stage in any protein blotting experiment and the resulting transfer should be a faithful replica of the original gel. The degree to which this is achieved is largely determined by the choice of transfer buffer, together with any pretransfer equilibration step, as well as the types of transfer media and transfer techniques that are used.

In the following sections two widely used methods of protein blotting—electroblotting (*3*) and passive-diffusion blotting (*4*)—are described. In addition, a method for staining proteins after transfer to nitrocellulose filters is given.

Materials

Unless otherwise indicated, all stock solutions and buffers are prepared from analytical grade reagents and are made up in deionized, double-distilled water. Buffers used in electroblotting experiments are stable for at least 3 months when stored at 4°C whereas passive-diffusion blotting buffers should be freshly prepared as required.

1. Equilibration buffer for electroblotting. The following chemicals are weighed out: 24.2 g Tris-base, 144.0 g glycine, and 10.0 g SDS (electrophoresis grade). They

are dissolved in approximately 7 L water, and to this solution 2 L methanol are added and the buffer is then made up to 10 L with water. The pH of this buffer is 8.3 and should not require any adjustment.

2. Transfer buffer for electroblotting. This buffer is identical to the equilibration buffer above, except that SDS should not be added (*see* Note 4).
3. Electroblotting equipment. A number of different types of transfer apparatus are now commercially available. Alternatively, equally satisfactory transfer assemblies can be made, without too much difficulty, in a laboratory workshop. One of the best designs for such an apparatus is described in a paper by Bittner et al. (*5*). The basic design given in this paper can be improved by substituting stiff nylon netting for the dialysis membrane used on the inner faces of the cassette frame to support the gel and filter paper "sandwich" (*see* Method). Also the efficiency of the packing can be further improved by using two Scotch Brite pads as additional packing (*see* Method) and rubber bands, instead of the screws, can be used to hold the transfer assembly together.
4. Equilibration buffer for passive-diffusion blotting. The following solutions are added to approximately 350 mL of water: 5mL 1.0*M* Tris-HCl, pH 7.0; 5 mL 5.0*M* NaCl; 10 mL 0.1*M* Na_2EDTA and 0.5 mL 0.1*M* dithiothreitol, followed by the addition of 120.12 g of urea ("Aristar" grade or similar). The solution is then mixed until all of the urea has dissolved and made up to a final volume of 500 mL.
5. Transfer buffer for passive-diffusion blotting. This buffer is identical to the equilibration buffer above, except that it contains no urea and a total of 4 L of buffer is required.
6. Passive-diffusion blotting equipment. As yet, no commercial transfer apparatus is available for passive-diffusion blotting experiments. However, the limited amount of apparatus required for this type of transfer can be easily made in a laboratory workshop (*see* Fig. 1).
7. Transfer filters. Nitrocellulose filters are used for both electro- and passive-diffusion blotting experiments and

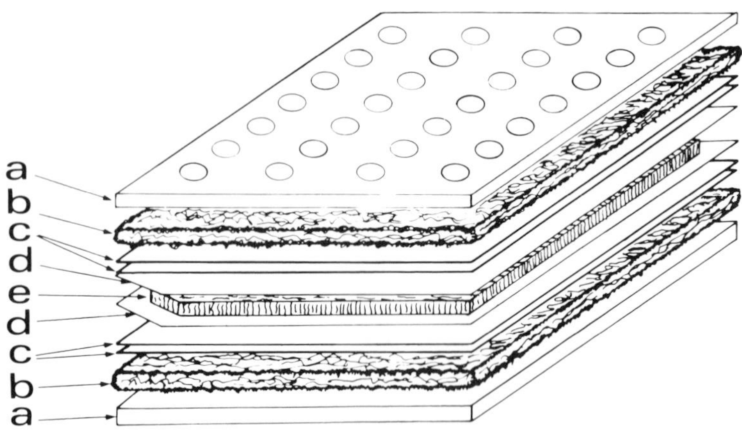

Fig. 1. Schematic representation of a transfer assembly for protein blotting by the passive-diffusion method. (a) rigid plastic sheets (20 × 20 cm, rows of 1-cm diameter holes at 2 cm intervals are drilled through these sheets in order to allow the free passage of transfer buffer), (b) Scotch Brite pads; (c) filter papers; (d) nitrocellulose filters and (e) the gel. **N.B.** The front left-hand corner has been cut off both the gel and the nitrocellulose filters in order to assist in the later orientation of the transferred proteins.

can be obtained from a number of suppliers. The majority of these suppliers produce filters with a 0.45 μm pore size and these are the most widely quoted in the literature. However, in common with a number of other workers, we prefer to use the smaller 0.2-μm pore size filters because of their greater binding capacity. Nitrocellulose filters should be stored in an airtight box at 4°C when not in use to prevent them from becoming contaminated by exposure to volatile chemicals, etc. Several other types of filters can be used in protein blotting experiments and these are briefly discussed in Note 3 below.

8. Amido black protein stain. About 30 min before the stain is required, a 1 L solution containing 20% (v/v) ethanol and 7% (v/v) acetic acid is prepared. Into 100 mL of this solution 0.1 g of Amido Black 10 B is dissolved by constant stirring for ~ 30 min, the remaining 900 mL of solution being held over for the destaining step.

Method

N.B. Disposable plastic gloves should be worn when handling either the gels or transfer filters in order to prevent them from being contaminated by grease and other substances.

Electroblotting

1. Proteins are separated by SDS polyacrylamide slab gel electrophoresis (SDS PAGE; *see* Chapter 6). A single sample can be run on the gel where a variety of probes or binding conditions are to be studied. Alternatively, when only a limited number of probes and/or binding conditions are under investigation, a large number of different samples may be analyzed.
2. At the end of the run, the position of the sample tracks is marked on both gel plates. The plates are then carefully separated using a large spatula as a lever, while a second smaller spatula is eased between the gel and one of the gel plates. (The separation of the gel plates can be further simplified by routinely siliconizing one of the plates.) The comb and stacking gel (if any) are then removed and in the case of single sample gels side strips are cut. Alternatively, where several sets of samples have been run on the gel, these are cut into separate strips. Polyacrylamide gels can be easily cut into strips with a large rounded scapel blade. In order to prevent the gel from tearing, the scalpel blade is slowly brought across the gel by advancing with a rocking action of the blade.
3. Before removing the gel strips from the plate one corner should be cut off each strip so as to assist in their later orientation.
4. The side strips or duplicate gels are then stained in Coomassie brilliant blue R 250 or investigated by any other suitable detection system (*see* Chapters 6 and 17)
5. The remaining gel strips are placed in a large glass dish containing approximately 750 mL of equilibration buffer. The gel is then gently agitated in this solution on a rotary shaker table for 60 min.

6. Open the transfer cassette, and place it in a large glass dish. The cassette is then assembled by layering the following, starting from the anode side: one Scotch Brite pad, two sheets of thick chromatography paper (e.g., Whatman 3MM grade), nitrocellulose filter, the gel to be blotted, followed by two further sheets of filter paper and a final Scotch Brite pad. The filter papers and the nitrocellulose filter should be thoroughly saturated with transfer buffer before being placed in the cassette. Uneven wetting of the nitrocellulose filters frequently presents a problem at this stage. However, this problem can be easily overcome by floating the filter across the surface of a dish containing transfer buffer. After a few seconds, the filter rapidly loses its pure white color, turning a translucent gray-white color. Once the filter is uniformly saturated with buffer, the dish is gently rocked, submerging the filter and ensuring its complete saturation. The Scotch Brite pads are very porous, quickly taking up the buffer when submerged in the transfer chamber, and consequently do not require any pre-wetting. As the transfer cassette is assembled, it is vital that no air bubbles are trapped between either the filters or the gel, otherwise an uneven transfer will result. The transfer cassette is then closed and secured with rubber bands (if required).
7. The assembled cassette is placed in the transfer chamber containing transfer buffer and lifted in and out of the buffer a few times to ensure that no air bubbles are trapped in the Scotch Brite pads. The lid of the transfer chamber is then put in place and the electrodes connected (the right way round!).
8. The transfer is performed at a current of ~0.5 amps (~10–15 V/cm) for between 2 and 20 h at 4°C. The precise length of the transfer is dependent upon a number of factors, including buffer composition, the percentage of acrylamide in the gel, and the molecular weight distribution of the proteins (high molecular weight proteins transferring more slowly). The optimum transfer time should therefore be determined by running a series of trial experiments, and monitoring the residual proteins in the gel as well as the proteins that have been transferred to the filter. (The transfer buffer can

be used for five transfers before it needs to be replaced.)

Passive-Diffusion Blotting

1. Proteins are separated by SDS PAGE: *see* Steps 1–4 above.
2. The gel is then placed in a large glass dish containing 500 mL of equilibration buffer and gently agitated for 3 h on a rotary shaker table.
3. The transfer cassette is assembled essentially as described above (Step 6) except that two nitrocellulose filters are used, one on each side of the gel (*see* Fig. 1).
4. The assembled transfer cassette is submerged in approximately 1.5 L of transfer buffer. After 24 h the buffer is replaced with fresh buffer and the transfer is allowed to continue for a further 2–4 d. (Again the optimum transfer time is dependent upon the composition of the gel and transfer buffer, as well as on the size and solubility of the proteins under investigation, and should ideally be established for each different class of experiments).

Protein Detection

1. Proteins can be detected by gently agitating the filter in the Amido black stain prepared previously (*see* Materials) for about 10 min.
2. The stain is then poured off and the filter is briefly rinsed with destain solution, followed by further washes in this solution, for approximately 1 h.
3. The filter is then dried between several sheets of filter paper, that are held flat with a heavy weight.

Notes

1. The potential applications of protein blotting are very wide-ranging and beyond the scope of this chapter. Examples of possible applications can be found in the following papers: immunoblotting (*6,7*), DNA binding

proteins (*8,9*), glycoproteins (*10, 11*), and receptor proteins (*12*).

2. The two methods that have been described in this chapter are both well-characterized and there are many references to their use in the literature. However, there are a number of alternative transfer techniques, such as Southern blotting (*13*) and vacuum blotting (*14*), but while these methods may be preferred for some applications they are not yet widely used.
3. Although nitrocellulose filters are the most commonly used filters in protein blotting experiments, several other types of transfer media are currently available.

   (a) Nylon filters. These filters are reported to have similar binding properties to nitrocellulose, but they have the advantage of being considerably stronger. Although the fragility of nitrocellulose filters is seldom a serious problem, the nylon-based filters may be valuable where the filters have to be handled a lot, for example, in some immunoblotting experiments where the filters are recycled. However, for the majority of applications the extra expense of these filters cannot be justified.

   (b) Diazo papers. A range of diazo papers, including both diazobenzyloxymethyl (DBM) cellulose and diazophenylthioether (DPT) cellulose papers, are now commerically available in their more stable intermediate amino forms. Alternatively, these papers can be prepared in the laboratory (*see* refs. *15* and *16* for the preparation of DBM and DPT papers, respectively). These papers are particularly useful where the transferred proteins need to be very tightly bound to the filter (as in some immunoblotting experiments) or possibly in facilitated active transfer experiments (*17*). Again few experiments currently justify the extra expense and time involved in preparing these filters.

   Both ABM and APT papers have short half-lives and they should only be used in electroblotting experiments. Also because the reactive diazo groups will bind not only proteins, but free amino acids, the Tris/glycine/methanol electroblotting buffer should be replaced with either a phosphate (*5*) or borate buffer (*18*).

(c) Diethylaminoethyl (DEAE) anion exchange papers and membranes. These filters are available from a number of suppliers and have been successfully employed in DNA blotting experiments. The major advantage of these filters is that they allow the recovery of the transferred molecules (*19*).

4. It is important to note that, using the method described above, it is unlikely that all of the proteins will be transferred from the gel. The explanation for this would apppear to be that there is preferential loss of SDS from the gel. As long as the SDS remains in the gel, the proteins are readily solubilized. However, owing to its small size and negative charge, the SDS migrates out of the gel more rapidly than the majority of the proteins, which are then trapped in the gel. The inclusion of SDS in the equilibration buffer considerably improves the transfer efficiency of this system. If SDS is also included in the transfer buffer, protein transfer reaches almost 100%. Unfortunately many proteins then fail to bind to the nitrocellulose filter, presumably because of the detergent action of SDS, which disrupts the hydrophobic interactions responsible for binding the proteins to the filter (*see* Fig. 2). Substituting nonionic detergents only aggravates this situation, the uncharged detergent doing little to improve the migration of the proteins out of the gel while at the same time preventing many proteins from binding to the nitrocellulose.
5. Low molecular weight proteins (~60,000 daltons or less) are generally transferred far more efficiently than larger proteins, irrespective of the composition of the transfer buffer. Gibson has recently published a simple method designed to improve the transfer of high molecular weight proteins (*20*). Proteins are electrophoretically transferred essentially as described above. The major difference is that a filter paper soaked in a nonspecific protease is placed on the cathode side of the gel. Once the transfer starts, the protease migrates into the gel digesting any protein molecules it encounters. The resulting peptides are then easily transferred

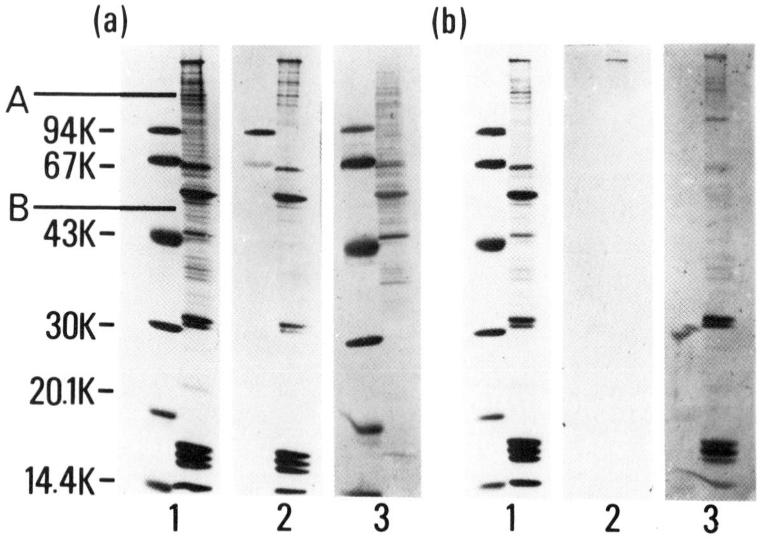

Fig. 2. Electroblot transfer efficiency. Pharmacia low molecular weight marker mix proteins and Chinese hamster nuclear proteins were separated by SDS polyacrylamide gel electrophoresis. The gels were then equilibrated in either (a) 20 m*M* Tris base, 192 m*M* glycine, 20% methanol or (b) the same buffer but containing 0.1% SDS, for 30 min. One gel from each buffer was then stained (*1*), while the remaining gels were electroblotted for 18.5 h at 0.5 A and 4°C, using the same buffers as used in the equilibration step. The post-transfer gels (2) and the filters (3) were then stained. [Polypeptides A and B, although minor components of the total nuclear protein sample, bind DNA probes with high affinity and in preference to all other nonhistone proteins (*see* Fig. 4, ref. *2*).]

out of the gel. The principal drawback of this approach is that potentially important sites within the proteins may be destroyed and will not be recognized by probes specific to these sites.

6. Passive-diffusion blotting transfer efficiency is approximately one-tenth to one-hundredth of that seen in electroblot transfers. Two major factors are responsible for this: (i) The transfer is a passive process as opposed to the active transfer that occurs in electroblotting experiments. (ii) The long equilibration step prior to the transfer removes most of the SDS from the gel, leaving

the less soluble proteins trapped in the gel. If the length of the equilibration step is decreased, the transfer efficiency is improved. However, the rationale in using such a long equilibration step is that the proteins have sufficient time to renature. These renatured proteins may then be expected to be more efficient than electrophoretically transferred proteins in binding the probe molecules. This increased binding efficiency should therefore, in part at least, counterbalance the low transfer efficiency.

7. Proteins from both acid and isoelectric focusing (IEF) gels can be electrophoretically transferred to nitrocellulose filters in 0.7% (v/v) acetic acid (*3*). Proteins in this case are transferred towards the *cathode* and the position of the nitrocellulose filter should be reversed with respect to the gel therefore. Alternatively proteins can be transferred from these gels following equilibration in electroblot transfer buffer containing 2.0% (w/v) SDS (store at room temperature), or for IEF gels by capillary blotting, as described by Reinhart and Malamud (*21*).
8. The detection of transferred proteins is an important step in any protein blotting experiment, providing both a qualitative and a quantitative measure of the transfer efficiency. The staining of the transferred proteins also makes it easier to interpret the eventual probe binding pattern. A number of alternative stain recipes for both Amido black and Coomassie brilliant blue are to be found in the literature. However, we have been unable to find any method that produces results superior to those obtained using the method described above. Indeed some of these methods use very high concentrations of ethanol or methanol, which can severely distort or even dissolve the filter! When radioactively labeled proteins are used, these can be detected either by fluorography for ^{3}H, ^{14}C, and ^{35}S using Enhance spray (New England Nuclear), or in the case of ^{125}I by autoradiography. These methods have the considerable advantage of increasing the sensitivity by several orders of magnitude and consequently are very valuable when determining the optimum transfer conditions for a particular blotting experiment. However, care must be taken when using labeled proteins to ensure

that they will not interfere with the detection of the probe, which is also often radioactively labeled. As an alternative to using radiolabeled proteins, a high sensitivity silver staining method has recently been developed for use on protein blots (22). Although in principal this is a very useful technique, the method is unfortunately very time-consuming.

References

1. Snabes, M. C., Boyd III, A. E., and Bryan, J. (1981) Detection of actin-binding proteins in human platelets by ^{125} I actin overlay of polyacrylamide gels. *J. Cell Biol.* **90,** 809–812.
2. Gooderham, K. (1983) Protein blotting. In *Techniques in molecular biology,* Walker J. M., and Gaastra, W., eds. pp 49–61. Croom Helm Publishers.
3. Towbin, H., Staehelin, T., and Gordon, J. (1979) Electrophoretic transfer of proteins from polyacrylamide gels to nitrocellulose: Procedure and some applications. *Proc. Natl. Acad. Sci. USA.* **76,** 4350–4354.
4. Bowen, B., Steinberg, J., Laemmli, U.K., and Weintraub, H. (1980) The detection of DNA-binding proteins by protein blotting. *Nucl. Acids Res.* **8,** 1–20.
5. Bittner, M., Kupferer, P., and Morris, C. F. (1980) Electrophoretic transfer of proteins and nucleic acids from slab gels to diazobenzyloxymethyl cellulose or nitrocellulose sheets. *Anal. Biochem.* **102,** 459–471.
6. Tsang, V. C. W., Peralta, J. M., and Simons, A. R. (1983) Enzyme-linked immunoelectrotransfer blot techniques (EITB) for studying the specificities of antigens and antibodies separated by gel electrophoresis. In *Methods in Enzymology* **92,** Langone, J. J. and Van Vunakis, H. V. eds. pp. 377–391. Academic Press, New York.
7. Burnette, W. N. (1981) "Western blotting"; Electrophoretic transfer of proteins from sodium dodecylsulfate-polyacrylamide gels to unmodified nitrocellulose and radiographic detection with antibody and radioiodinated Protein A. *Anal. Biochem.* **112,** 195–203.
8. Jack, R. S., Brown, M. T., and Gehring, W. J. (1983) Protein blotting as a means to detect sequence-specific DNA-

binding proteins. *Cold Spring Harbor Symp. Quant. Biol.*, **XLV11,** 483–491.

9. Triadou, P., Crepin, M., Gros, F., and Lelong, J.-C. (1982) Tissue-specific binding of total and β-globin genomic deoxyribounucleic acid to nonhistone chromosomal proteins from mouse erythropoietic cells. *Biochem.* **21,** 6060–6065.
10. Glass II, W. F., Briggs, R. C., and Hnilica, L. S. (1981) Use of lectins for detection of electrophoretically separated glycoproteins transferred onto nitrocellulose sheets. *Anal. Biochem.* **115,** 219–224.
11. Hawkes, R. (1982) Identification of concavalin A-binding proteins after sodium dodecyl sulfate-gel electrophoresis and protein blotting. *Anal. Biochem.* **123,** 143–146.
12. Oblas, B. Boyd, N. D., and Singer, R. H. (1983) Analysis of receptor-ligand interactions using nitrocellulose gel transfer: Application to *Torpedo* acetylcholine receptor and alpha-bungarotoxin. *Anal. Biochem.* **130,** 1–8.
13. Southern, E. M. (1975) Detection of specific sequences among DNA fragments separated by gel electrophoresis. *J. Mol. Biol.* **98,** 503–517.
14. Peferoen, M., Huybrechts, R., De Loof, A. (1982) Vacuum-blotting: a new simple and efficient transfer of proteins from sodium dodecyl sulfate-polyacrylamide gels to nitrocellulose. *FEBS Lett.* **145,** 369–372.
15. Alwine, J. C., Kemp, D. J., and Stark, G. R. (1977) Method for detection of specfic RNAs in agarose gels by transfer to diazobenzyloxymethyl-paper and hybridisation with DNA probes. *Proc. Natl. Acad. Sci. USA* **74,** 5350–5354.
16. Seed, B. (1982) Diazotizable arylamine cellulose paper for the coupling and hybridisation of nucleic acids. *Nucl. Acids Res.* **10,** 1799–1810.
17. Erlich, H. A., Levinson, J. R. Cohen, S.N., and McDevitt, H. O. (1979) Filter affinity transfer. A new technique for the in situ identification of proteins in gels. *J. Biol. Chem.* **254,** 12,240–12,247.
18. Reiser, J. and Wardale, J. (1981) Immunological detection of specific proteins in total cell extracts by fractionation in gels and transfer to diazophenylthioether paper. *Eur. J. Biochem.* **114,** 569–575.
19. Danner, D. B. (1982) Recovery of DNA fragments from gels by transfer to DEAE-paper in an electrophoresis chamber. *Anal. Biochem.* **125,** 139–142.
20. Gibson, W. (1981) Protease-facilitated transfer of high-molecular-weight proteins during electrotransfer to nitrocellulose. *Anal. Biochem.* **118,** 1–3.

21. Reinhart, M. P., and Malamud, D. (1982) Protein transfer from isoelectric focusing gels: The native blot. *Anal. Biochem.* **123,** 229–235.
22. Yuen, K. C. L. (1982) A silver-staining technique for detecting minute quantities of proteins on nitrocellulose paper: Retention of antigenicity of stained proteins. *Anal. Biochem.* **126,** 398–402.

Chapter 21

Peptide Mapping by Thin-Layer Chromatography and High Voltage Electrophoresis

Keith Gooderham

MRC Clinical and Population Cytogenetics Unit, Western General Hospital, Crewe Road, Edinburgh, United Kingdom

Introduction

Thin-layer chromatography and electrophoresis, either separately or in combination, provide a simple, high resolution technique for producing peptide maps. Recently, thin-layer methods have tended to be increasingly overshadowed by the development of high performance liquid chromatography (HPLC). Although thin-layer methods are generally not as sensitive as HPLC (requiring milli- or microgram amounts of protein instead of the nanogram or less quantities used in HPLC), they are still preferred for many applications. Thin-layer peptide mapping not only has the advantage of being relatively inexpensive, but it is also a very simple technique and it is often possible to analyze a large number of different samples in a single experiment. In addition, a number of different stains can be used for the detection of specific amino acids and these considerably increase the value of this method.

Although thin-layer methods are probably most widely used in comparative peptide mapping experiments (*see,* for example, Figs. 1 and 2), they also find a variety of other applications including: (i) monitoring enzyme and chemical cleavages of proteins; (ii) the preparative isolation of peptides for microsequencing studies (*1*); (iii) assaying peptide purity following preparative paper chromatography/electrophoresis, ion-exchange and gel filtration chromatography, and so on—a single band on a thin-layer plate, together with a single *N*-terminal amino acid, usually being a good indication that the sample contains only one peptide.

Two different methods are used in thin-layer peptide mapping experiments; namely high voltage electrophoresis and chromatography. In high voltage electrophoresis experiments, peptides are fractionated primarily upon the basis of charge, whereas the hydrophobic properties of the peptides are more important for chromatographic separations. Both methods may be used independently to produce one-dimensional peptide maps or alternatively these methods may be combined to produce two-dimensional peptide maps. In the latter case, it is usually necessary to first produce a series of one-dimensional peptide maps in order to establish the optimum separation conditions for the peptides under investigation. Once these conditions have been established, they should be rigorously adhered to in a given series of experiments, in order to ensure that comparable peptide maps are obtained, and ideally all of the samples should be prepared at the same time and as far as possible worked up in parallel.

Materials

1. Preprepared thin-layer cellulose plates (20 cm × 20 cm × 0.1 mm) can be obtained from a number of suppliers.

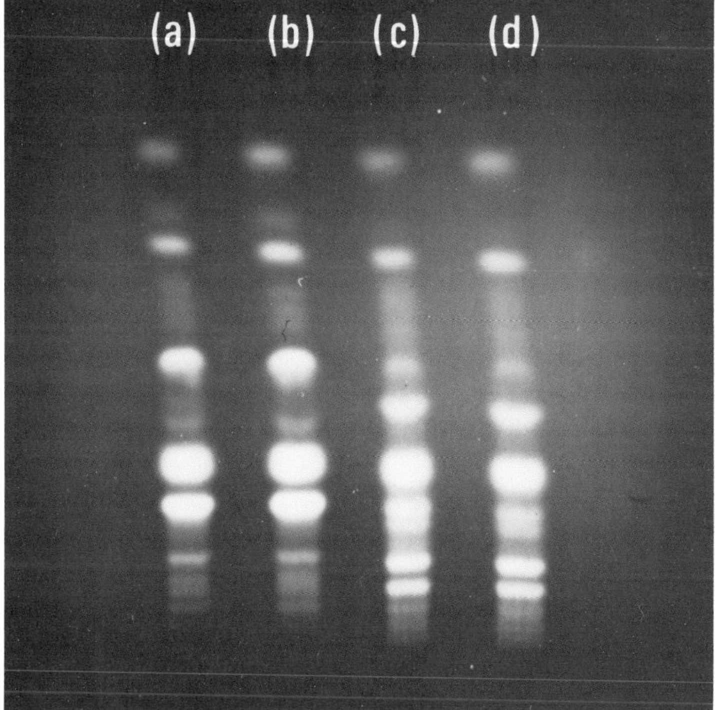

Fig. 1. Comparative one-dimensional tryptic peptide maps of pig and calf thymus HMG 1 and HMG 2 chromosomal proteins after staining with phenanthraquinone. Protein samples (a) pig HMG 1; (b) calf HMG 1; (c) pig HMG 2; and (d) calf HMG 2 at 5 mg/mL in 0.2*M* ammonium bicarbonate solution were digested with trypsin at an enzyme to substrate ratio of 1:50 (w/w) for 48 h at 37°C. Peptides were then separated by ascending thin-layer chromatography in a mixture of butan-1-ol, acetic acid, water and pyridine (15:3:10:12, by volume) and arginine containing peptides were detected by staining with phenanthraquinone. HMG 1 and 2 proteins are two very closely related chromosomal proteins [~80% homology (*11*)]. However, when tryptic peptides from these proteins are separated by one-dimensional thin-layer chromatography followed by specific staining for arginine-containing peptides the two proteins are clearly shown to be different.

For thin-layer chromatography, either glass or aluminum supports are equally acceptable, but for thin-layer high voltage electrophoresis only glass-backed plates can be used.

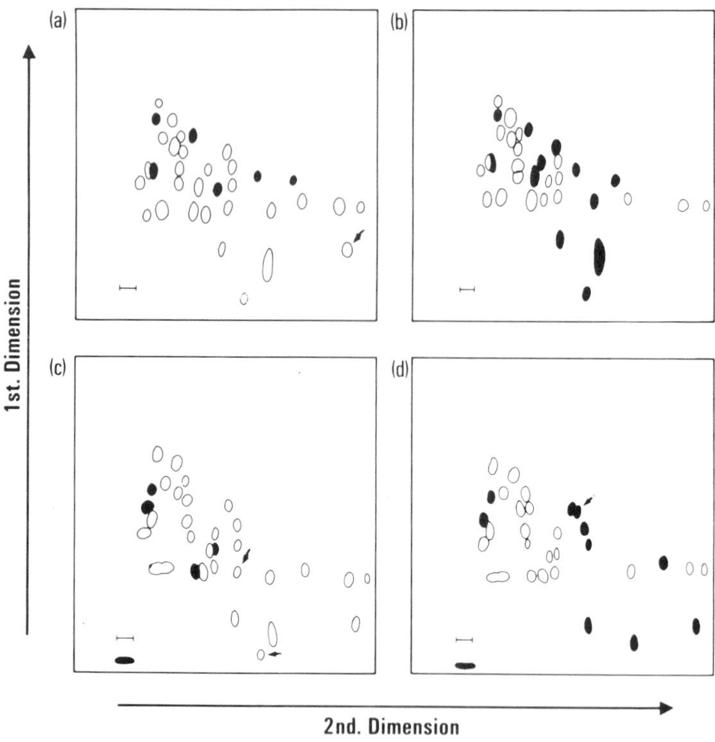

Fig. 2. Comparative two-dimensional tryptic peptide maps of pig and calf thymus HMG 1 and HMG 2 chromosomal proteins. Protein samples (a) pig HMG 1; (b) calf HMG 1; (c) pig HMG 2; and (d) calf HMG 2 at 5 mg/mL in 0.2*M* ammonium bicarbonate solution were digested with trypsin at an enzyme to substrate ratio of 1:50 (w/w) for 48 h at 37°C. Peptides were then separated by thin-layer high voltage electrophoresis in pH 6.5 buffer at 2.5 kV for 25 min (1st dimension). After drying the thin-layer plate the peptides were separated (2nd dimension) by ascending chromatography in a mixture of butan-1-ol, acetic acid, water, and pyridine (15:3:10:12, by volume). The peptides were detected by staining with ninhydrin–cadmium acetate solution and their positions were recorded on tracing paper. Although the primary sequences of HMG 1 and 2 proteins are very similar [80% homologous (*11*)] when the peptide maps for these proteins are examined [i.e., compare (a) with (c) and (b) with (d)] marked differences between these proteins are observed (peptides unique to either HMG 1 or 2 are shown as solid spots). Similarly, although the primary sequences of these proteins are clearly very conserved [compare (a) and (b) and also (c) and (d)] species specific differences (shown by arrows) are clearly apparent using this method (c.f., Fig. 1).

2. Chromatography tanks capable of accepting two or more plates can be obtained through most laboratory supply houses. Ideally these tanks should be made of glass in order to avoid any possible reaction between the tank and the various solvents which will be used. In order to ensure an air-tight seal between the tank and the lid, the rim of the tank should be lightly coated with silicone grease.
3. High voltage electrophoresis apparatus and power supplies are available from a number of suppliers. Probably the most versatile of these is the Model L24 and associated power supply manufactured by Shandon Southern Instruments. All of the procedures described below are intended for use with this instrument, but they should be equally suitable for use with other instruments with little or no modification.
4. Aerosol spray guns can be obtained from most laboratory supply houses.
5. Wherever possible, electrophoresis buffers are prepared with analytical grade reagents and distilled water is used throughout. All of these buffers should be stored at room temperature and they can be kept for up to 6 months without any apparent deterioration. Some suitable electrophoresis buffers are listed below.

   (a) pH 6.5 buffer is prepared by mixing together pyridine, acetic acid, and water in the following proportions 25:1:225 by volume.
   (b) pH 4.8 buffer is prepared by mixing together pyridine, acetic acid, and water in the following proportions 3:3:394 by volume.
   (c) pH 3.5 buffer is prepared by mixing together pyridine, acetic acid, and water in the following proportions 1:10:189 by volume.
   (d) pH 2.0 buffer is prepared by mixing together acetic acid, formic acid (98% v/v), and water in the following proportions 8:2:90 by volume.

6. Wherever possible, all chromatography solvents are prepared with analytical grade reagents and distilled water is used throughout. All solutions should be stored at room temperature and they can be kept for up to 6 months without any apparent deterioration. Some suitable chromatography solvents are listed below:

(a) Butan-1-ol:acetic acid:water:pyridine ("BAWP") are mixed together in the following proportions 15:3:10:12 by volume.
(b) Butan-1-ol:acetic acid:water are mixed together in a separating funnel in the following proportions 4:1:5 by volume. After allowing the phases to separate, the lower phase is poured off and discarded while the upper phase is retained.
(c) Butan-1-ol:urea:acetic acid are mixed together in the following proportions 4:5:1 (v/w/v). (This solution should be freshly prepared for each experiment.)

7. For preparing stains, all stock solutions are prepared with analytical grade reagents wherever possible and distilled water is used throughout. Unless otherwise indicated, all solutions should be stored in the dark and can be kept for up to 6 months. Some commonly used stains are listed below. Stains (a) to (d) are general purpose stains whereas stains (e) to (g) are stains for specific amino acids:

(a) Ninhydrin cadmium acetate (*2*). Two stock solutions are prepared:
Solution A. 6 g of cadmium acetate is dissolved in a mixture of 600 mL of water and 300 mL acetic acid.
Solution B. A 0.5% (w/v) solution of ninhydrin in acetone is prepared.
The stain is made by mixing 15 mL of Solution A with 85 mL of Solution B and is used immediately.
(b) Ninhydrin butan-1-ol 0.5% (w/v) solution of ninhydrin in butan-1-ol is prepared.
(c) Fluorescamine (*3*). A 0.001% (w/v) solution of fluorescamine in a mixture of acetone and pyridine (99:1 v/v) is prepared.
(d) *o*-Phthaldehyde (OPA) (*4*). A 3 g quantity of boric acid is dissolved in approximately 90 mL of water and adjusted to pH 10.5 with concentrated potassium hydroxide solution. To this solution, 0.05 g of OPA dissolved in 1 mL of ethanol plus 0.05 mL of 2-mercaptoethanol, is added. The solutions are then mixed and made up to 100 mL with water. Finally 0.3 mL of a 30% (w/v) solution of Brij is added.

(e) Ehrlich stain for tryptophan (5). A 1% (w/v) solution of *p*-dimethylaminobenzaldehyde in acetone containing 10% (v/v) hydrochloric acid is prepared. This solution is then stored at 4°C and can be kept for up to 2 months.

(f) Pauly stain for histidine and tyrosine (5). The following solutions are prepared as required.

Solution C. 5% (w/v) sodium nitrite.

Solution D. 1% (w/v) sulfanilic acid in 1*M* hydrochloric acid.

Solution E. 15% (w/v) sodium carbonate in water.

(g) Phenanthraquinone stain for arginine (6). The following solutions are prepared:

Solution F. A 0.02% (w/v) phenanthrenequinone solution in ethanol is prepared. This solution should be stored in the dark and can be kept for up to 3 months.

Solution G. A 10% (w/v) NaOH solution in 60% (v/v) ethanol is prepared just before the stain is to be used.

The stain is then prepared by mixing equal volumes of Solutions F and G and is used immediately.

Method

N.B. Disposable plastic gloves should be worn in all of the subsequent steps, not only to prevent the thin-layer plates from becoming contaminated, but also as a protection against the various stains and solvents used during the course of this work. A fume hood should also be used wherever possible and particularly when drying or spraying plates.

Tryptic Digest

1. The protein sample is dissolved at a concentration of between 2 and 10 mg/mL in 0.2*M* ammonium bicarbonate. A freshly prepared solution of trypsin (DCC-treated, bovine trypsin) at a concentration of 1 mg/mL in water is then added to this solution, giving a final enzyme-to-substrate ratio of 1:50.

2. The sample is then mixed and incubated at 37°C for 48 h.
3. The resulting digest is either stored at −20°C until required or taken straight on to the next step.

Thin-Layer Electrophoresis

1. An origin line is carefully drawn across the center of the plate using a soft pencil. The samples are then spotted on to the plate using a 10 or 25 μL syringe as 1-cm long bands along this line, leaving a 1.5–2.0 cm gap between each sample. Ideally the syringe needle should never touch the surface of the plate or the cellulose thin layer will be damaged. Where large sample volumes (i.e., greater than 10 μL) have to be loaded, a hair drier can be used to accelerate the drying of the sample.
2. Once the samples are dry the plate should be uniformly saturated with electrophoresis buffer, applied as an aerosol from a spray gun. If the surface of the plate becomes too wet at this stage, i.e., if pools of buffer form on the surface of the plate, the excess buffer should be carefully blotted off with a sheet of filter paper.
3. The plate is then placed, sample side uppermost, on top of a sheet of cellulose acetate covering the cooling platen of the electrophoresis apparatus. The electrode wicks (Whatman No. 1 filter paper or similar) saturated in electrophoresis buffer are then positioned so as to cover about 1 cm of the top and bottom edge of the plate. Again excess buffer is blotted off with spare sheets of filter paper. (If there is too much free buffer present at this stage, a short-circuit is likely to occur once the run starts).
4. The upper insulating sheet of cellulose acetate is then placed on top of the filter paper wicks and the thin-layer plate and the lid of the chamber closed. In a preliminary experiment only a very short run is required, ~10–15 min at 2 kV.
5. The plate is then removed and dried in a stream of warm air in a fume hood and the peptide separations are examined after staining with one of the nonspecific stains described below.
6. Having determined the basic pattern of the peptide separation it is then possible to optimize the separation

by moving the origin towards either the anode or cathode as appropriate and increasing the length of the run proportionally. Alternatively, should the first buffer system fail to give a satisfactory separation, one of the other buffer systems given in the Materials section should be tried. (The electrophoresis buffer should be replaced at least every 2 weeks.)

Thin-Layer Chromatography

1. An origin line is drawn as for high voltage electrophoresis, but instead of being across the middle of the plate the line should be 2 cm from one edge of the plate. The samples are then loaded as described previously.
2. The plate is then placed in a chromatography tank containing sufficient solvent to cover about 1 cm of the plate, but not enough to risk covering the samples or they will be washed off. The chromatography is then allowed to proceed, usually overnight, until the solvent reaches the top of the plate. (The solvent in the chromatography tank should be replaced every 2 weeks in order to ensure that the best separations are obtained.)
3. At the end of the run the chromatogram is removed from the tank and dried in a stream of warm air in a fume hood and stained.

Two-Dimensional Peptide Mapping

Having established the optimum conditions for both the electrophoretic and the chromatographic separation of the peptides, it is then possible to extend the analysis to two dimensions. Although this means that only one sample can be analyzed per plate it is frequently justified in terms of the increased resolution which can be obtained (compare Figs. 1 and 2).

The procedure for producing two-dimensional peptide maps is essentially as described in the previous two sections. Theoretically either electrophoresis or chromatography may be used for the first dimension separation, but generally better results are obtained where electrophoresis is used for the first dimension and chromatography in the second dimension.

1. The sample is loaded along a 1 cm origin line, parallel to the bottom of the plate and 2 cm from the left-hand side of the plate. The precise position of the origin line from the bottom of the plate should be determined by a trial one-dimensional separation as described above.
2. After running the sample in the first dimension the plate is thoroughly dried in a stream of warm air.
3. The plate is then rotated through 90° with the former left-hand edge of the plate now forming the bottom of the plate and the peptides are separated by ascending chromatography as described previously.

Diagonal Peptide Mapping

Two-dimensional thin-layer peptide mapping experiments rely on a combination of two different fractionation techniques in order to obtain the best possible separation of peptides. If instead of using two different separation methods, the same system is used in both the first and second dimensions the peptides will form a single diagonal line across the plate. Although this diagonal separation does not normally offer any advantage over a simple one-dimensional separation it can be exploited for the identification of cysteine containing peptides (*7*):

1. The sample is loaded on to a thin-layer plate and the peptides are then separated by thin-layer high voltage electrophoresis or chromatography as described previously.
2. At the end of the run the plate is thoroughly dried in a stream of warm air.
3. While the plate is drying, a solution of performic acid is prepared by mixing together 95 mL of 98% (v/v) formic acid and 5 mL of hydrogen peroxide. This solution is then placed in a shallow dish, which is placed at the bottom of a chromatography tank.
4. The dry plate is placed in this tank and left for 2 h, after which time the plate is removed from the tank and carefully dried once more.
5. The thin-layer plate is then rotated through 90°, i.e., the former left-hand edge of the plate becomes the new

bottom edge of the plate and the peptides are run in the second dimension, using the same separation system as before.

6. At the end of the run the plate is again dried and stained (*see* below). The majority of the peptides will be seen to lie along a single diagonal line. However, peptides that were previously joined together by S—S bridges will now have different mobilities owing to the intervening performic acid oxidation step and the resulting cysteic acid containing peptides will fall off this line.

Stains

All of the following stains (*see* also Chapter 4) are applied as aerosols from a distance of 30–50 cm, ensuring that the plate is uniformly saturated with reagent. It is important, however, that the plates do not become too wet or the peptide separations will become blurred.

(a) Ninhydrin cadmium acetate. After spraying the plate it is left to dry for 5 min before being placed in an oven at 110°C for 5–10 min. The majority of the peptides will appear as red spots against a pale pink background. However, peptides with either glycine or threonine *N*-terminal amino acids will initially appear as yellow spots that gradually turn red over the course of 1–2 d, while peptides with an *N*-terminal valine frequently stain weakly if at all.

(b) Ninhydrin butan-1-ol. Peptides are detected as described in (a) above and appear as blue or purple spots against a light purple background. (The lack of contrast between the nonspecific background staining and the peptides can make these separations more difficult to photograph in comparison to peptides stained with ninhydrin–cadmium acetate reagent). Unlike the ninhydrin–cadmium acetate reagent this stain also has the disadvantage of reacting with the side chains of lysine residues that may give a false impression of the abundance of peptides containing a large number of lysine residues.

(c) Fluorescamine. After spraying the plate it is left to dry for 15–20 min (if the plate is too wet much of the fluorescence will be quenched) before examining the plate under a long wavelength (366 nm) ultraviolet light. As fluorescamine only reacts with primary amines peptides with an *N*-terminal proline residue will not be detected, but these peptides can be deteced by staining with ninhydrin once the fluorescamine staining pattern has been recorded. The failure to detect proline residues is, however, more than comepensated for by the fivefold increase in sensitivity compared to that obtained with ninhydrin.

(d) *o*-Phthaldehyde. After spraying the plate it is left to dry for 15–20 min (again if the plate is too wet much of the fluorescence will be quenched) before examining the plate under a long wave length (366 nm) ultraviolet light. Peptides are visible by their intense blue fluorescence (visible for up to 2 h) and as little as 10 pmol of a single peptide may be detected, making this by far the most sensitive of the general purpose stains.

(e) Ehrlich reagent for tryptophan. After spraying the plate it is allowed to dry and tryptophan-containing peptides are seen as bright purple spots that gradually fade over a number of hours. This stain may be used on a plate that has previously been stained with ninhydrin as the hydrochloric acid in the Ehrlich's reagent will bleach the ninhydrin-stained peptides.

(f) Pauly reagent for histidine and tyrosine. Solutions C and D are left on ice for 20 min. Equal volumes of these two solutions are then mixed together and the plate is immediately sprayed with this solution. After allowing the plate to dry (~20 min) the plate is then sprayed with Solution E and again left to dry. Histidine and tyrosine-containing peptides will appear as either red or weak brown spots, respectively.

(g) Phenanthraquinone stain for arginine. After staining the plate is thoroughly dried (~20 min). Arginine-containing peptides may then be visualized by examining the plate under a long wavelength (366 nm) ultraviolet light; the stained peptides will remain brightly fluorescent for at least 18 h. After recording the distribution of these peptides they can be stained with either ninhydrin–cadmium acetate solution or with the Pauly reagent if required. In the case of the Pauly reagent, the plate will already be sufficiently basic following the phenanthraquinone staining to permit the Pauly reagent to be used without needing to use Solution E.

Notes

1. Although trypsin is a very effective protease it will only cleave those peptide bonds that are exposed on the surface of the protein. In proteins that contain disulfide bridges some of these potential cleavage sites will be buried within the protein molecule and consequently will be protected from the trypsin. In proteins where this is a problem this can be overcome by reducing the cystines and then blocking the SH groups by S-carboxymethylation as described by Crestfield et al. (*8*). (*See* also p. 35.)
2. In the majority of peptide mapping experiments peptides are produced by digestion with trypsin. This enzyme is a specific protease, which cleaves peptide bonds on the C-terminal side of arginine and lysine residues. These amino acids are relatively abundant in most proteins, accounting for approximately 11.5% of the total amino acids in an "average" protein (*9*). However, a number of other specific enzymatic and chemical cleavages are available (for examples, *see 10*) and these can be successfully used where the amino acid or sequence analysis for a particular protein indicates that they will produce a better range of peptides.
3. The choice of which electrophoresis buffer to use is largely empirical. Peptides containing large numbers of acidic and basic residues are usually best resolved in the more basic buffers, while good separations of peptides rich in neutral amino acids are generally obtained with the more acidic buffer systems.
4. In the majority of chromatography experiments the "BAWP" solvent will produce excellent results, however, should this prove unsatisfactory, one of the other solvent systems should be tried.

5. A number of the stains described in this chapter are now commercially available, in ready to use, aerosol cans. Further details about these stains can be found in the catalogs of many of the leading chemical manufactures (*see also* Chapter 4 for other available stains).

References

1. Powers, D. A., Fishbein, J. C., and Place, A. R. (1983). Thin-layer peptide mapping with sequencing at the nanomole level. In *Methods in Enzymology* **91,** Hirs, C. H. W. and Timasheff, S. N. eds. pp. 466–486. Academic Press, New York.
2. Heilman, J., Barrollier, J., and Watzke, E. (1957) Beitrag zur aminosaurebestimmung auf papierchromatogrammen. *Hoppe-Seylers Zeit.* **309,** 219–220.
3. Felix, A. M., and Jimenez, M. H. (1974) Usage of fluorescamine as a spray reagent for thin-layer chromatography. *J. Chromatogr.* **89,** 361–364.
4. Gardner, W. S., and Miller III, W. H. (1976) Reverse-phase liquid chromatographic analysis of amino acids after reaction with o-pathaladehyde. *Anal. Biochem.* **101,** 61–65.
5. Bennett, J. C. (1967) Paper chromatography and electrophoresis; special procedure for peptide maps. In *Methods in Enzymology,* XI, *Enzyme structure,* pp. 330–339. (Hirs, C. H. W. ed.) Academic Press, New York.
6. Yamada, S., and Itano, H. A. (1966) Phenanthrenequinone as an analytical reagent for arginine and other monosubstituted guanidines. *Biochim. Biophys. Acta* **130,** 538–540.
7. Brown, J. R., and Hartley B. S. (1966) Location of disulphide bridges by diagonal paper electrophoresis. The disulphide bridges of bovine chymotrypsinogen A. *Biochem. J.* **101,** 214–228.
8. Crestfield, A. M., Moore, S., and Stein, W. H. (1963) The preparation and enzymatic hydrolysis of reduced and S-carboxymethlated proteins. *J. Biol. Chem.* **238,** 622–627.
9. Dayhoff, M. O., Hunt, L. T., and Hurst-Calderone, S. (1978) *Atlas of Protein Sequence and Structure* **5,** Supplement 3, pp. 363–373 (ed. Dayhoff, M. O.) National Biomedical Research Foundation, Washington, DC.
10. Croft, L. R. (1980) *Handbook of Protein Sequence Analysis.* 2nd edition, Wiley, New York.
11. Walker, J. M., Gooderham, K., Hastings, J. R. B., Mayes, E., and Johns, E. W. (1980) The primary structures of nonhistone chromosomal proteins HMG 1 and 2. *FEBS Lett.* **122,** 264–270.

Chapter 22

In Situ Peptide Mapping of Proteins Following Polyacrylamide Gel Electrophoresis

Keith Gooderham

MRC Clinical and Population Cytogenetics Unit, Western General Hospital, Crewe Road, Edinburgh, United Kingdom

Introduction

Polyacrylamide gel electrophoresis is a simple, yet versatile, high resolution technique for the analysis of complex mixtures of proteins. However, this is not to say that this method is without problems. For example, where a series of different protein samples are run on a gel, many of the proteins will have the same mobility and it is frequently impossible to be certain if these bands represent the same protein or whether they simply share similar mobilities. Conversely the samples may contain degradation products or structurally related proteins of differing mobilities.

In this chapter a simple *in situ* peptide-mapping technique, based on a method originally developed by Cleveland et al. (*1*) is described. Briefly the method is as follows: Proteins are separated by discontinuous sodium dodecyl sulfate polyacrylamide slab gel electrophoresis (SDS PAGE) and stained with Coomassie brilliant blue (*see* Chapter 6). After destaining, bands containing polypeptides of interest are cut out of the gel and equilibrated against Tris pH 6.8 buffer. The gel slices are then ready for re-electrophoresis and together with a suitable protease are loaded onto a second SDS polyacrylamide slab gel and run into the stacking gel at a low current. As the proteins and protease migrate through the low percentage polyacrylamide pH 6.8 stacking gel, the proteins are digested by the protease and the resulting peptides are then resolved in the lower pH 8.8 separating gel.

Peptides are then detected by standard methods of staining, autoradiography, and so on. The recent development of high-sensitivity silver staining methods for the detection of proteins following polyacrylamide gel electrophoresis (*see* Chapter 13) has proved to be a very valuable method when used in conjunction with the peptide-mapping technique described here. Silver staining methods are between 20 and 100 times more sensitive than traditional Coomassie staining methods (*see* Fig. 1), while not significantly less sensitive than fluorography following labeling with tritium. Silver staining also has the advantage of not requiring any special sample preparation, and the actual staining procedure takes appproximately the same amount of work as is involved in preparing a gel for fluorography.

In situ peptide mapping therefore requires very little extra work or equipment and yet considerably extends the information potential of both one- and two-dimensional polyacrylamide gels. There are a large number of possible applications of this method, the most obvious and important of which are: (i) comparison of comigrating proteins (*see,* for example, Fig. 1); (ii) identification of structurally related proteins; (iii) identification of degradation products; and (iv) detection of different domains, e.g., antigen-binding sites, following protein blotting (*see* Chapter 20). In addition, although this method has been most widely used for the analysis of proteins separated by SDS PAGE, it is a very versatile technique, and with suitable modifications, it can be used with almost any combination of gel systems (*2*).

Materials

1. Gel apparatus. Both the primary protein separation and the secondary peptide-mapping gels are run in a vertical slab gel apparatus based on the design originally described by Studier (*3*). The spacers and comb for the primary gel are 1 mm thick and a 14-well comb with 1.5 cm deep and 1 cm wide wells is used. The spacers and combs for the secondary gel are 1.5 mm thick and a 20-well comb with 6 mm wide and 20 mm deep wells is used. The increased thickness of the secondary gel is necessary in order to accommodate the gel slices that swell considerably during the staining and equilibration steps. Similarly the increased depth of the secondary gel wells is required in order to accommodate the gel slices while leaving sufficient room for the addition of the protease buffer.
2. Equilibration buffer. The following solutions are mixed together: 1.25*M* Tris-HCl, pH 6.8 (10 mL), 10% SDS (1 mL, electrophoresis grade), 0.1*M* Na_2EDTA, pH 7.0 (1 mL) and made up to a final volume of 100 mL with water.
3. Overlay buffer. The following solutions are mixed together: 1.25*M* Tris-HCl, pH 6.8 (10 mL), 10% SDS (1 mL, electrophoresis grade), 0.1*M* Na_2EDTA, pH 7.0 (1 mL), glycerol (20 mL), 1%(w/v) Bromophenol blue (1 mL) and made up to a final volume of 100 mL with water.
4. Protease buffer. The following solutions are mixed together: 1.25*M* Tris-HCl, pH 6.8 (10 mL), 10% SDS (1 mL, electrophoresis grade), 0.1*M* Na_2EDTA, pH 7.0 (1 mL), glycerol (10 mL), 1% (w/v) Bromophenol blue (1 mL) and made up to a final volume of 100 mL with water.
5. Proteases. A number of different proteases have been used for peptide mapping experiments with SDS

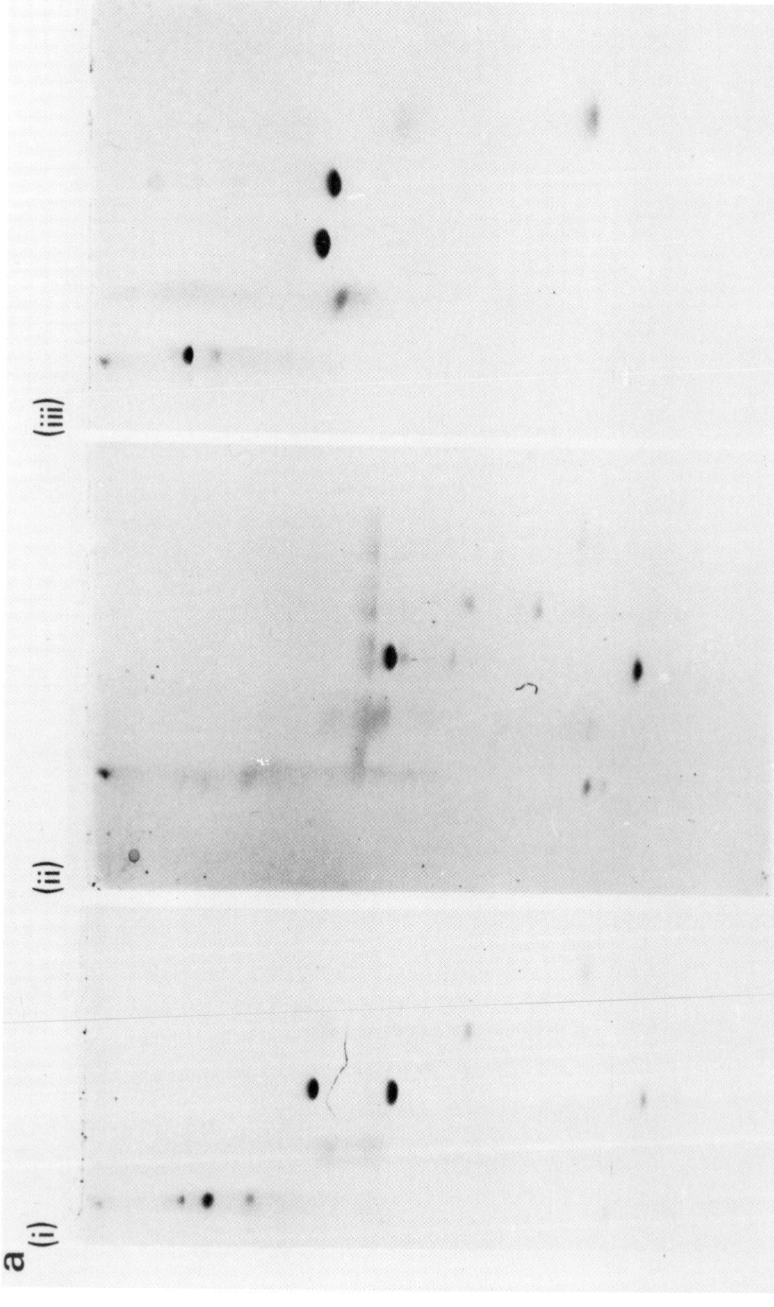
a
(i)
(ii)
(iii)

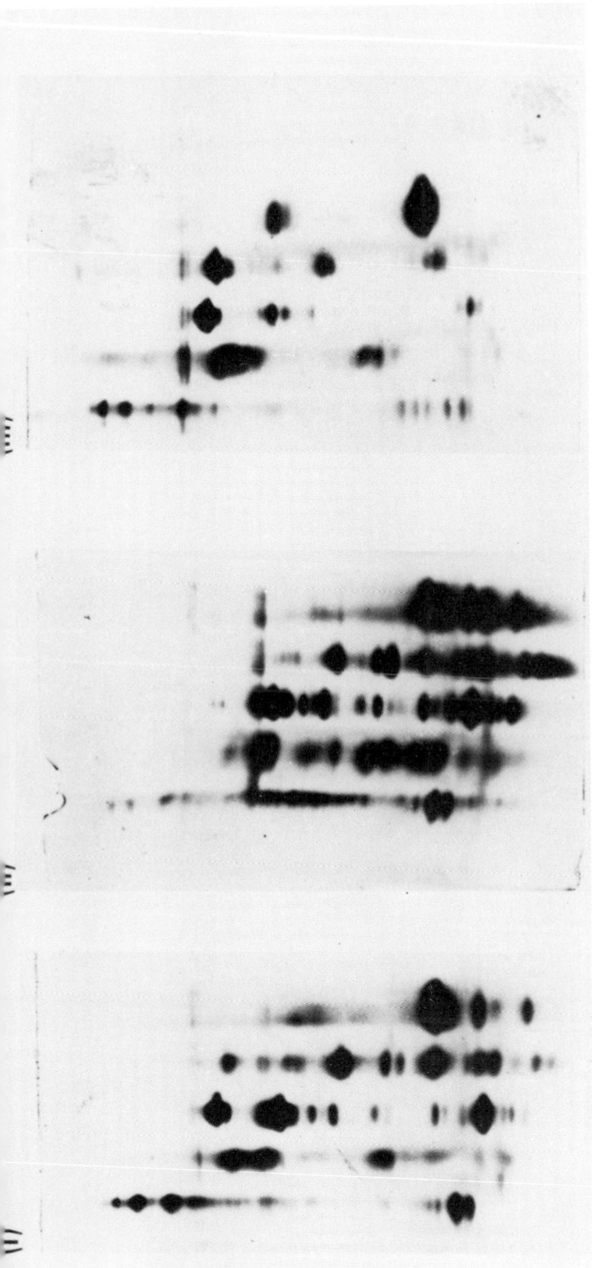

Fig. 1. In situ peptide maps of five standard proteins. The effect of protease concentration and a comparison of Coomassie and silver stains. Six peptide maps, each containing an identical panel of proteins (from left to right: phosphorylase a, 100,000 d (daltons); ovalbumin, 43,000 d; 3-phosphoglyceric phosphokinase, 43,000 d; actin, 43,000 d; and β-lactoglobulin, 35,000 d. 2.5 μg of each protein was loaded) are shown. The proteins were initially run on two 15% (w/v) polyacrylamide SDS gels (not shown). After staining with Coomassie brilliant blue the individual protein bands were cut out and loaded onto two 1.5 mm thick peptide mapping gels [15% (w/v) polyacrylamide] and subjected to *in situ* peptide mapping with (i) 0.05μg V8 protease; (ii) 0.5 μg V8 protease; and (iii) no protease. The resulting peptides were then visualized by either (a) Coomassie brilliant blue or (b) by silver staining (*see* Chapter 13).

polyacrylamide gels. Of these, the most frequently used are *Staphylococcus aureus* V8 protease, chymotrypsin, and papain. V8 protease is relatively expensive and is most conveniently prepared as a 1 mg/mL stock solution in water and stored in aliquots at −20°C. Each aliquot may be frozen and thawed several times without any apparent loss of activity. The other two enzymes are freshly prepared as 1 mg/mL stock solutions for each experiment. Where other gel systems are to be used, e.g., acid gels, it may not be possible to use these enzymes, but either pepsin or one of the chemical cleavage techniques (*see* Note 6) may well prove to be suitable alternatives.

Unless otherwise indicated, all stock solutions and buffers are prepared from analytical grade reagents and are made up in deionized double-distilled water. With the exception of the SDS and Bromophenol blue stock solutions, which are stored at room temperature, all other solutions and buffers are stored at 4°C and are stable for at least three months.

Method

1. Proteins are separated by polyacrylamide slab gel electrophoresis (SDS PAGE or any other suitable system), followed by staining with Coomassie brilliant blue R 250 (Chapter 6).
2. The gel should then be photographed in order to produce a permanent record of the proteins under investigation. Bands containing proteins of interest are then cut out of the gel and placed in a 10 mL screw-top plastic test tube containing 5 mL of equilibration buffer. The tubes are then laid on a shaker table and gently agitated. After 1 h the buffer is replaced with a further 5 mL of fresh buffer and the equilibration continued for a further hour.
3. The buffer is then drained off and the gel slices can then conveniently be stored at −20°C until required.
4. After thawing, the gel slices are placed (short-end down) in the sample wells of an extra thick (1.5 mm) SDS polyacrylamide gel with a 5 cm stacking gel. (This is twice the normal size of stacking gel, but is necessary

in order to allow sufficient time for the protease to digest the proteins, as well as to permit the efficient "stacking" of the samples.)

5. The gel slices are covered with overlay buffer followed by 10 μL of protease buffer containing 0.05 μg of protease.
6. The samples are then run at a constant current of 20 mA until the marker dye reaches the top of the separating gel when the current is increased to 40 mA for the remainder of the run. The total run time is about 6 h.
7. Peptides are then stained overnight in Coomassie brilliant blue R250 or fixed and subsequently visualized by silver staining (Chapter 13), fluorography (Chapter 17), and so on.

Notes

1. The versatility of the peptide mapping technique makes it suitable for use with a wide range of different gel systems. The peptide mapping gel can be either the same or different to the one used for the initial protein separation, providing that a suitable intervening equilibration step is included. However, for most applications the SDS PAGE system (*4*) is compatible with the widest range of enzymes, as well as offering the best resolution of the resulting peptides. The percentage of acrylamide used in the separating gel is determined by the molecular weight of the proteins under investigation as well as by the size of the resulting peptides. As a rough guide, an 8% (w/v) gel should be used for proteins of 90,000 daltons or greater, whereas for smaller proteins either 15% (w/v) or 20% (w/v) gels will produce the best results. Alternatively, where a wide range of different sized peptides are generated a gradient gel can be used (*5*).
2. In Cleveland's original paper (*1*), a very brief staining step (15 min) is used prior to the removal of the protein bands and their being carried on to the equilibration step. However, this is not necessary and the proteins can be completely stained before being removed from the gel, with the result that minor proteins can be detected more readily. This also allows the protein sepa-

ration to be photographed and therefore avoids the need to run a duplicate gel. However, by exposing the proteins to the very acidic staining and destaining solutions for several days, peptide bonds between aspartic and proline residues may be hydrolyzed [*see* Fig. 1a (iii) and 1b (iii)]. The extent to which this occurs depends upon a number of factors, including the length of the staining and destaining steps, as well as the temperature at which the gel is stored. Therefore, in order to obtain comparable peptide maps, it is important that all of the proteins should, wherever possible, be taken from the same gel, or where this is not practical, the gels should be processed in exactly the same way.

3. Once the bands containing the proteins of interest have been cut out of the gel they can be stored at −20°C immediately, or at the end of the first stage of the equilibration step, or upon completing this step. The gel slices should be carefully drained before freezing or they may disintegrate when thawed. Once frozen, the samples can be stored for as long as three months.
4. In the methods section above, an enzyme concentration of 0.05 μg/sample is recommended. At this concentration all three enzymes will produce acceptable peptide maps over a wide range of protein concentrations (0.1–5 μg), assuming that the protein contains suitable cleavage sites. However, in some cases, it may be worthwhile to try a range of different enzyme concentrations in order to obtain a better distribution of peptides.
5. When a series of different gel slices contain varying amounts of protein, this can be overcome by initially running a trial gel and measuring the concentration of the proteins either by densitometry (Chapter 14) or alternatively by overnight extraction of the Coomassie stained protein with 25% (v/v) pyridine. The optical densities of these extracts are then measured at 595 nm (*6*) and the sample volumes adjusted accordingly so as to produce bands of equal intensity. However, this method will generally tolerate quite a wide range of enzyme and substrate concentrations and any differences that do occur tend to be quantitative rather than qualitative (*see* Fig. 1).

6. In addition to using the various proteolytic enzymes described above, a number of chemical cleavage methods have been described. These include *N*-chlorosuccinimide (*7*), cyanogen bromide (*8*), hydroxylamine (*8*), and formic acid (*9*). All of these methods involve rather extreme reaction conditions that are unlikely to be compatible with the gel system used for the peptide separation. The proteins are therefore cleaved in the gel slices and prior to the equilibration step. The sole purpose of the secondary gel is then to separate the peptides and consequently the size of the stacking gel can be reduced to 2.5 cm while proportionally increasing the size of the separating gel.
7. Where marker proteins are required, either to identify the starting protein or as molecular weight standards, they should be taken from the same gel as the other proteins. If this is not done and they are simply loaded directly onto the peptide mapping gel in solution they will travel faster than the other proteins. This is because these proteins are delayed in their migration by first having to travel out of the gel slices before entering the mapping gel.
8. A wide range of methods has been used for detecting peptides produced by this technique and they all have their respective advantages and disadvantages. Generally, Coomassie brilliant blue is not sufficiently sensitive (*see* Fig. 1) and we have favored the use of one of the high-sensitivity silver stains (*see* Fig. 1 and Chapter 13). The major disadvantage of this technique is that it is nonspecific, staining not only the peptides, but also the protease. Much cleaner results can be obtained using either radioactively labeled proteins or, alternatively, proteins that have been labeled with dansyl chloride, but both of these require advance planning and preparation.

References

1. Cleveland, D. W., Fischer, S. G., Kirschner, M. W., and Laemmli, U. K. (1977) Peptide mapping by limited proteolysis in sodium dodecyl sulfate and analysis by gel electrophoresis. *J. Biol. Chem.* **252,** 1102–1106.

2. Spiker, S. (1980) Slab gel designed for enzymatic digestion of proteins in polyacrylamide gel slices and direct resolution of peptides. *J. Chromatog.* **198,** 169–171.
3. Studier, W. F. (1973) Analysis of bacteriophage T7 early RNAs and proteins on slab gels. *J. Mol. Biol.* **79,** 237–248.
4. Laemmli, U. K. (1970) Cleavage of structural proteins during the assembly of the head of bacteriophage T4. *Nature* **227,** 680–685.
5. Tijssen, P., and Kurstak, E. (1979) A simple and sensitive method for the purification and peptide mapping of proteins solubilized from densonucleosis virus with sodium dodecyl sulfate. *Anal. Biochem.* **99,** 97–104.
6. Fenner, C., Traut, R. R., Mason, D. T., and Wilkeman-Coffelt, J. (1975) Quantification of Coomassie blue stained proteins in polyacrylamide gels based on analyses of eluted dyes. *Anal. Biochem.* **63,** 595–602.
7. Lischwe, M. A., and Ochs, D. (1982) A new method for partial peptide mapping using *N*-chlorosuccinimide/urea and peptide silver staining in sodium dodecyl sulfate polyacrylamide gels. *Anal. Biochem.* **127,** 453–457.
8. Lam, K. S., and Kasper, C. B. (1979) Electrophoretic analysis of three major nuclear envelope polypeptides. Topological relationship and sequence homology. *J. Biol. Chem.* **254,** 11,713–11,720.
9. Sonderegger, P., Jaussi, R., Gehring, H., Brunschweiler, K., and Christen, P. (1982) Peptide mapping of protein bands from polyacrylamide gel electrophoresis by chemical cleavage in gel pieces and re-electrophoresis. *Anal. Biochem.* **122,** 298–301.

Chapter 23

The Dansyl Method for Identifying *N*-Terminal Amino Acids

John M. Walker

School of Biological and Environmental Sciences, The Hatfield Polytechnic, Hatfield, Hertfordshire, England

Introduction

The reagent 1-dimethylaminonaphthalene-5-sulfonyl chloride (dansyl chloride, DNS-Cl) reacts with the free amino groups of peptides and proteins as shown in Fig. 1. Total acid hydrolysis of the substituted peptide or protein yields a mixture of free amino acids plus the dansyl derivative of the *N*-terminal amino acid, the bond between the dansyl group and the *N*-terminal amino acid being resistant to acid hydrolysis. The dansyl amino acid is fluorescent under UV light and is identified by thin-layer chromatography on polyamide sheets. This is an extremely sensitive method for identifying amino acids and in partic-

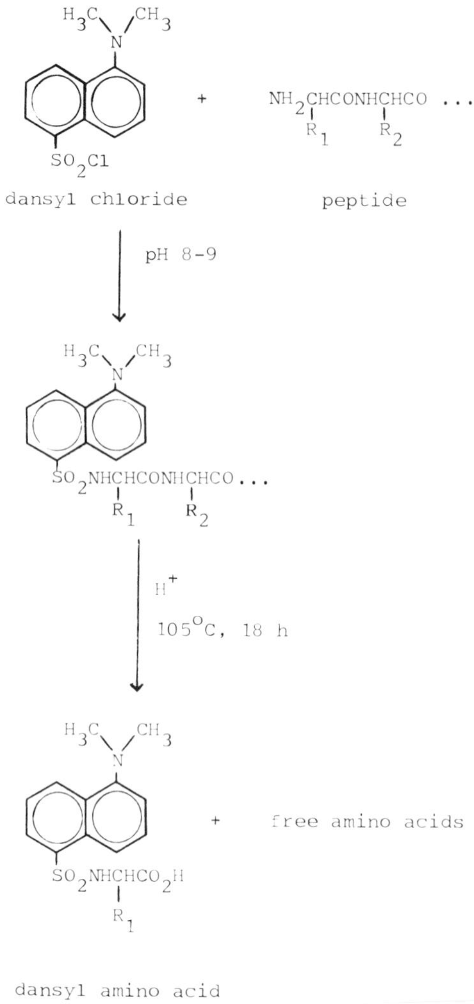

$H_3C \backslash N / CH_3$ (dansyl chloride: naphthalene ring, SO_2Cl) + $NH_2CHCONHCHCO$ ... (R_1, R_2)

dansyl chloride peptide

pH 8-9

$SO_2NHCHCONHCHCO$... (R_1, R_2)

H^+

$105^{\circ}C$, 18 h

$SO_2NHCHCO_2H$ (R_1) + free amino acids

dansyl amino acid

Fig. 1. Reaction sequence for the labeling of *N*-terminal amino acids with dansyl chloride.

ular has found considerable use in peptide sequence determination when used in conjunction with the Edman degradation (*see* Chapter 24). The dansyl technique was originally introduced by Gray and Hartley (*1*), and was developed essentially for use with peptides. However, the method can also be applied to proteins (*see* Note No. 12).

Materials

1. Dansyl chloride solution (2.5 mg/mL in acetone). Store at 4°C in the dark. This sample is stable for many months. The solution should be prepared from concentrated dansyl chloride solutions (in acetone) that are commercially available. Dansyl chloride available as a solid invariably contains some hydrolyzed material (dansyl hydroxide).
2. Sodium bicarbonate solution (0.2*M*, aqueous). Store at 4°C. Stable indefinitely, but check periodically for signs of microbial growth.
3. 5*N* HCl (aqueous).
4. Test tubes (50 × 6 mm) referred to as "dansyl tubes."
5. Polyamide thin layer plates (7.5 × 7.5 cm). These plates are coated on both sides, and referred to as "dansyl plates." Each plate should be numbered with a pencil in the top corner of the plate. The origin for loading should be marked with a pencil 1 cm in from each edge in the lower left-hand corner of the numbered side of the plate. The origin for loading on the reverse side of the plate should be immediately behind the loading position for the front of the plate, i.e., 1 cm in from each edge in the lower *right-hand* corner.
6. Three chromatography solvents are used in this method.

   Solvent 1: Formic acid:water, 1.5:100, v/v
   Solvent 2: Benzene:acetic acid, 9:1, v/v
   Solvent 3: Ethyl acetate:methanol:acetic acid, 2:1:1, v/v/v

7. An acetone solution containing the following standard dansyl amino acids. Pro, Leu, Phe, Thr, Glu, Arg (each approximately 50 μg/mL).
8. A UV source, either long wave (265 μm) or short wave (254 μm).

Method

1. Dissolve the sample to be analyzed in an appropriate volume of water, transfer to a dansyl tube, and dry *in*

vacuo to leave a film of peptide (1–5 nmol) in the bottom of the tube.

2. Dissolve the dried peptide in sodium bicarbonate (0.2*M*, 10 μL) and then add dansyl chloride solution (10 μL) and mix.
3. Seal the tube with parafilm and incubate at 37°C for 1 h, or at room temperature for 3 h.
4. Dry the sample *in vacuo*. Because of the small volume of liquid present, this will only take about 5 min.
5. Add 6*N* HCl (50 μL) to the sample, seal the tube in an oxygen flame, and place at 105°C overnight (18 h).
6. When the tube has cooled, open the top of the tube using a glass knife, and dry the sample *in vacuo*. If phosphorus pentoxide is present in the desiccator as a drying agent, and the desiccator is placed in a water bath at 50–60°C, drying should take about 30 min.
7. Dissolve the dried sample in 50% pyridine (10 μL) and, using a microsyringe, load 1 μL aliquots at the origin on *each side* of a polyamide plate. This is best done in a stream of warm air. Do not allow the diameter of the spot to exceed 3–4 mm.
8. On the *reverse side only*, also load 0.5 μL of the standard mixture at the origin.
9. When the loaded samples are completely dry, the plate is placed in the first chromatography solvent and allowed to develop until the solvent front is about 1 cm from the top of the plate. This takes about 10 min, but can vary depending on room temperature.
10. Dry both sides of the plate by placing it in a stream of warm air. This can take 5–10 min since one is evaporating an aqueous solvent.
11. If the plate is now viewed under UV light, a blue fluorescent streak will be seen spreading up the plate from the origin, and also some green fluorescent spots may be seen within this streak. However, no interpretations can be made at this stage.
12. The dansyl plate is now developed in the second solvent, at right angles to the direction of development in the first solvent. The plate is therefore placed in the chromatography solvent so that the blue 'streak' runs along the bottom edge of the plate.

13. The plate is now developed in the second solvent until the solvent front is about 1 cm from the top of the plate. This takes 10–15 min.
14. The plate is then dried in a stream of warm air. This will only take 2–3 min since the solvent is essentially organic. However, since benzene is involved, drying *must* be done in a fume cupboard.
15. The side of the plate containing the sample only should now be viewed under UV light. Three major fluorescent areas should be identified. Dansyl hydroxide (produced by hydrolysis of dansyl chloride) is seen as a blue fluorescent area at the bottom of the plate. Dansyl amide (produced by side reactions of dansyl chloride) has a blue–green fluorescence and is about one-third of the way up the plate. These two spots will be seen on all dansyl plates and serve as useful internal markers. Occasionally other marker spots are seen and these are described in the Notes section below. The third spot, which normally fluoresces green, will correspond to the dansyl derivative of the *N*-terminal amino acid of the peptide or protein. However, if the peptide is not pure, further dansyl derivatives will of course be seen. The separation of dansyl derivatives after solvent 2 is shown in Fig. 2. Solvent 2 essentially causes separation of the dansyl derivatives of hydrophobic and some neutral amino acids, whereas derivatives of charged and other neutral amino acids remain at the lower end of the chromatogram.
16. A reasonable identification of any faster-moving dansyl derivatives can be made after solvent 2 by comparing their positions, relative to the internal marker spots, with the diagram shown in Fig. 2. Unambiguous identification is made by turning the plate over and comparing the position of the derivative on this side with the standard samples that were also loaded on this side.

    N.B. Both sides of the plate are totally independent chromatograms. There is no suggestion that fluorescent spots can be seen *through* the plate from one side to the other.

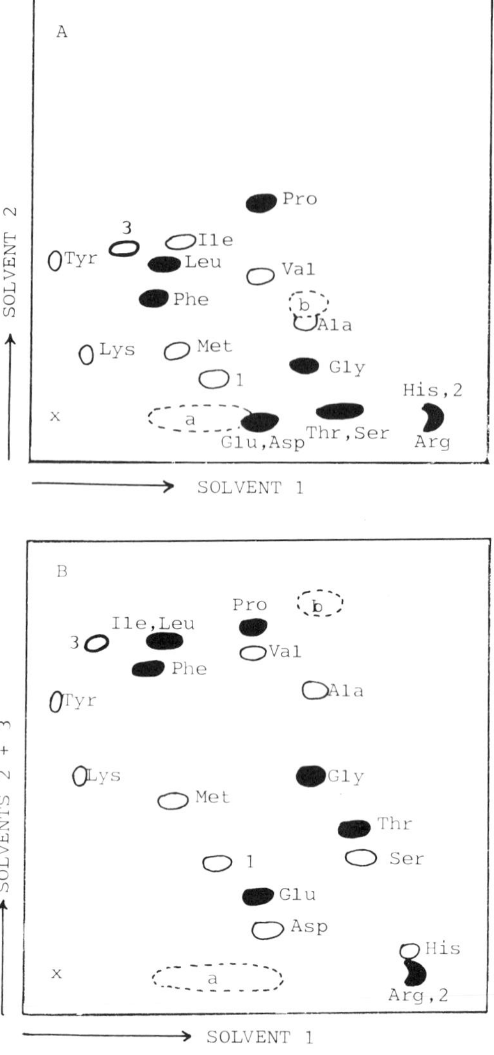

Fig. 2. Diagrams showing the separation of dansyl amino acids on polyamide plates after two solvents (A), and after three solvents (B): a = dansyl hydroxide; b = dansyl amide; 1 = tyrosine (*o*-DNS-derivative); 2 = lysine (ϵ-DNS-derivative); 3 = histidine (bis-DNS-derivative). The standard dansyl amino acids that are used are indicated as black spots.

17. Having recorded one's observations after the second solvent, the plate is now run in solvent 3 in the *same direction* as solvent 2. The plate is run until the solvent is 1 cm from the top, and this again takes 10–15 min.
18. After drying the plate in a stream of warm air (1–2 min), the plate is again viewed under UV light. The fast-running derivatives seen in solvent 2 have now run to the top of the plate and are generally indistinguishable (hence the need to record one's observations after solvent 2). However, the slow-moving derivatives in solvent 2 have now been separated by solvent 3 and can be identified if present. The separation obtained after solvent 3 is also shown in Fig. 2. The sum of the observations made after solvents 2 and 3 should identify the number and relative intensities of *N*-terminal amino acids present in the original sample.

Notes

1. It is important that the initial coupling reaction between dansyl chloride and the peptide occurs in the pH range 9.5–10.5. This pH provides a compromise between the unwanted effect of the aqueous hydrolysis of dansyl chloride and the necessity for the *N*-terminal amino group to be unprotonated for reaction with dansyl chloride. The condition used, 50% acetone in bicarbonate buffer, provides the necessary environment. The presence of buffer or salts in the peptide (or protein) sample should therefore be avoided to prevent altering the pH to a value outside the required range.
2. Because of the unpleasant and irritant nature of pyridine vapor, loading of samples onto the dansyl plates should preferably be carried out in a fume cupboard.
3. The viewing of dansyl plates under UV light should always be done wearing protective glasses or goggles. Failure to do so will result in a most painful (although temporary) conjunctivitis.
4. Most dansyl derivatives are recovered in high ($>90\%$) yield. However, some destruction of pro-

line, serine, and threonine residues occurs during acid hydrolysis, resulting in yields of approximately 25, 65, and 70%, respectively. When viewing these derivatives, therefore, their apparent intensities should be visually 'scaled-up' accordingly.

5. The sensitivity of the dansyl method is such that as little as 1–5 ng of a dansylated amino acid can be visualized on a chromatogram.
6. The side chains of both tyrosine and lysine residues also react with dansyl chloride. When these residues are present in a peptide (or protein), the chromatogram will show the additional spots, *o*-DNS-Tyr and ϵ-DNS-Lys, which can be regarded as additional internal marker spots. The positions of these residues are shown in Fig. 2. These spots should not be confused with bis-DNS-Lys and bis-DNS-Tyr, which are produced when either lysine or tyrosine is the *N*-terminal amino acid.
7. At the overnight hydrolysis step, dansyl derivatives of asparagine or glutamine are hydrolyzed to the corresponding aspartic or glutamic acid derivatives. Residues identified as DNS-Asp or DNS-Glu are therefore generally referred to as Asx or Glx, since the original nature of this residue (acid or amide) is not known. This is of little consequence if one is looking for a single *N*-terminal residue to confirm the purity of a peptide or protein. It does, however, cause difficulties in the dansyl-Edman method for peptide sequencing (*see* Chapter 24) where the residue has to be identified unambiguously.
8. When the first two residues in the peptide or protein are hydrophobic residues, a complication can occur. The peptide bond between these two residues is particularly (although not totally) resistant to acid hydrolysis. Under normal conditions, therefore, some dansyl derivative of the first amino acid is produced, together with some dansyl derivative of the *N*-terminal dipeptide. Such dipeptide derivatives generally run on chromatograms in the region of phenylalanine and valine. However, their behavior in solvents 2 and 3, and their positions relative to the marker derivatives should prevent misidentification

as phenylalanine or valine. Such dipeptide spots are also produced when the first residue is hydrophobic and the second residue is proline, and these dipeptide derivatives run in the region of proline. However, since some of the *N*-terminal derivative is always produced, there is no problem in identifying the *N*-terminal residue when this situation arises. A comprehensive description of the chromatographic behavior of dansyl–dipeptide derivatives has been produced (*2*).

9. Three residues are difficult to identify in the three solvent system described in the methods section; DNS-Arg and DNS-His because they are masked by the ϵ-DNS-Lys spot, and DNS-Cys because it is masked by DNS-hydroxide. If these residues are suspected, a fourth solvent is used. For arginine and histidine, the solvent is

   0.05*M* trisodium phosphate:ethanol (3:1, v/v)

   For cysteine the solvent is

   1*M* ammonia:ethanol (1:1, v/v)

   Both solvents are run in the same direction as solvents 2 and 3, and the residues are identified by comparison with relevant standards loaded on the reverse side of the plate.

10. When working with small peptides it is often of use also to carry out the procedure known as "double dansylation." Having identified the *N*-terminal residue, the remaining material in the dansyl tube is dried down and the dansylation process (steps 2–4) repeated. The sample is then redissolved in 50% pyridine (10 μL) and a 1 μL aliquot is examined chromatographically. The chromatogram will now reveal the dansyl derivative of *each* amino acid present in the peptide. Therefore for relatively small peptides ($<$ 10 residues) a quantitative estimation of the amino acid composition of the peptide can be obtained. This method is not suitable for larger peptides or proteins since most residues will be present more than once in this case, and it is not possible to

quantitatively differentiate spots of differing intensity.

11. Although the side chain DNS-derivative will be formed during dansylation if histidine is present in the peptide or protein sequence, this derivative is unstable to acid and is not seen during *N*-terminal analysis. Consequently, *N*-terminal histidine yields only the α-DNS derivative and not the bis-DNS compound as might be expected. The bis-DNS derivative *is* observed, however, if the mixture of free amino acids formed by acid hydrolysis of a histidine-containing peptide is dansylated and subsequently analyzed chromatographically (i.e., during "double dansylation").
12. The dansylation method described here was developed for use with peptides. However, this method can also be applied quite successfully to proteins, although some difficulties arise. These are caused mainly by insolubility problems, which can limit the amount of reaction between the dansyl chloride and protein thus resulting in a lower yield of dansyl derivative, and the presence of large amounts of *o*-DNS-Tyr and ε-DNS-Lys on the chromatogram that can mask DNS-Asp and DNS-Glu. A modification of the basic procedure described here for the dansylation of proteins has been described (*3*).

References

1. Gray, W. R., and Hartley, B. S. (1963) A fluorescent end group reagent for peptides and proteins. *Biochem. J.* **89,** 59P.
2. Sutton, M. R., and Bradshaw, R. A. (1978) Identification of dansyl dipeptides. *Anal. Biochem.* **88,** 344–346.
3. Gray, W. R. (1967) in *Methods in Enzymology* Vol. XI (ed. Hirs, C. H. W.) p. 149. Academic Press, New York.

Chapter 24

The Dansyl-Edman Method for Peptide Sequencing

John M. Walker

School of Biological and Environmental Sciences, The Hatfield Polytechnic, Hatfield, Hertfordshire, England

Introduction

The dansyl-Edman method for peptide sequencing uses the Edman degradation (*see* Chapter 26) to sequentially remove amino acids from the *N*-terminus of a peptide. Following the cleavage step of the Edman degradation, the thiazolinone derivative is extracted with an organic solvent *and discarded*. This contrasts with the direct Edman degradation method (Chapter 26), where the thiazolinone is collected, converted to the more stable PTH derivative, and identified. Instead, a small fraction (~5%) of the remaining peptide is taken and the newly liberated *N*-terminal amino acid determined in this sample by the dansyl method (*see* Chapter 23). Although the dansyl-Edman method results in successively less peptide being present at each cycle of the Edman degradation, this loss of material is more than compensated for by the fact that the dansyl method for identifying *N*-terminal amino acids is about one hundred times more sensitive than methods for identifying PTH amino acids. The dansyl-Edman method described here was originally introduced by Hartley (*1*).

Materials

1. Ground glass stoppered test tubes (approx. 65 × 10 mm, e.g., Quickfit MF 24/0). All reactions are carried out in this 'sequencing' tube.
2. 50% Pyridine (aqueous, made with AR pyridine). Store under nitrogen at 4°C in the dark. Some discoloration will occur with time, but this will not affect results.
3. Phenylisothiocyanate (5% v/v) in pyridine (AR). Store under nitrogen at 4°C in the dark. Some discoloration will occur with time, but this will not affect results. The phenylisothiocyanate should be of high purity and is best purchased as "sequenator grade." Make up fresh about once a month.
4. Water-saturated *n*-butyl acetate. Store at room temperature.
5. Anhydrous trifluoracetic acid (TFA). Store at room temperature under nitrogen.

Method

1. Dissolve the peptide to be sequenced in an appropriate volume of water, transfer to a sequencing tube, and dry *in vacuo* to leave a film of peptide in the bottom of the tube. (*see* Note 1.)
2. Dissolve the peptide in 50% pyridine (200 μL) and remove an aliquot (5 μL) for *N*-terminal analysis by the dansyl method (*see* Chapter 23).
3. Add 5% phenylisothiocyanate (100 μL) to the sequencing tube, mix gently, flush with nitrogen and incubate the stoppered tube at 50°C for 45 min.

4. Following this incubation, unstopper the tube and place it *in vacuo* for 30–40 min. The desiccator should contain a beaker of phosphorus pentoxide to act as a drying agent, and if possible the desiccator should be placed in a water bath at 50–60°C. When dry, a white 'crust' will be seen in the bottom of the tube. This completes the coupling reaction.
5. Add TFA (200 μL) to the test-tube, flush with nitrogen and incubate the stoppered tube at 50°C for 15 min.
6. Following incubation, place the test tube *in vacuo* for 5 min. TFA is a very volatile acid and evaporates rapidly. This completes the cleavage reaction.
7. 'Dissolve' the contents of the tube in water (200 μL). Do not worry if the material in the tube does not all appear to dissolve. Many of the side-products produced in the previous reactions will not in fact be soluble.
8. Add *n*-butyl acetate (1.5 mL) to the tube, mix vigorously for 10 s, and then centrifuge in a bench centrifuge for 3 min.
9. Taking care not to disturb the lower aqueous layer, carefully remove the upper organic layer and discard.
10. Repeat this butyl acetate extraction procedure once more and then place the test-tube containing the aqueous layer *in vacuo* (with the desiccator standing in a 60°C water bath if possible) until dry (30–40 min).
11. Redissolve the dried material in the test tube in 50% pyridine (200 μL) and remove an aliquot (5 μL) to determine the newly liberated *N*-terminal amino acid (the second one in the peptide sequence) by the dansyl method.
12. A further cycle of the Edman degradation can now be carried out by returning to step 3. Proceed in this manner until the peptide has been completely sequenced.

Notes

1. Manual sequencing is normally carried out on peptides between 2 and 30 residues in length and requires 1–5 nmol of peptide.

2. Since the manipulative procedures are relatively simple it is quite normal to carry out the sequencing procedure on 8 or 12 peptides at one time.
3. A repetitive yield of 90–95% is generally obtained for the dansyl-Edman degradation. Such repetitive yields usually allow the determination of sequences up to 15 residues in length, but in favorable circumstances somewhat longer sequences can be determined.
4. The amino acid sequence of the peptide is easily determined by identifying the new *N*-terminal amino acid produced after each cycle of the Edman degradation. However, because the Edman degradation does not result in 100% cleavage at each step, a background of *N*-terminal amino acids builds up as the number of cycles increases. Also, as sequencing proceeds, some fluorescent spots reflecting an accumulation of side products can be seen towards the top of the plates. For longer runs (10–20 cycles) this can cause some difficulty in identifying the newly liberated *N*-terminal amino acid. This problem is best overcome by placing the dansyl plates from consecutive cycles adjacent to one another and viewing them at the same time. By comparison with the previous plate, the increase of the new residue at each cycle, over and above the background spots, should be apparent.
5. A single cycle takes approximately 2.5 h to complete. When this method is being used routinely, it is quite easy to carry out three or four cycles on eight or more peptides in a normal day's work. During the incubation and drying steps the dansyl samples from the previous days sequencing can be identified.
6. As sequencing proceeds, it will generally be necessary to increase the amount of aliquot taken for dansylation at the beginning of each cycle, since the amount of peptide being sequenced is reduced at each cycle by this method. The amount to be taken should be determined by examination of the intensity of spots being seen on the dansyl plates.
7. It is most important that the sample is completely dry following the coupling reaction. Any traces of water present at the cleavage reaction step will introduce

hydrolytic conditions that will cause internal cleavages in the peptide and a corresponding increase in the background of *N*-terminal amino acids.

8. Very occasionally it will prove difficult to completely dry the peptide following the coupling reaction and the peptide appears oily. If this happens, add ethanol (100 μL) to the sample, mix, and place under vacuum. This should result in a dry sample.
9. When removing butyl acetate at the organic extraction step, take great care not to remove *any* of the aqueous layer as this will considerably reduce the amount of peptide available for sequencing. Leave a small layer of butyl acetate above the aqueous phase. This will quickly evaporate at the drying step.
10. Vacuum pumps used for this work should be protected by cold traps. Considerable quantities of volatile organic compounds and acids will be drawn into the pump if suitable precautions are not taken.
11. The identification of *N*-terminal amino acids by the dansyl method is essentially as described in Chapter 23. However, certain observations peculiar to the dansyl-Edman method are described below.
12. Tryptophan cannot be identified by the dansyl method as it is destroyed at the acid hydrolysis step. However, where there is tryptophan present in the sequence, an intense purple color is seen at the cleavage (TFA) step involving the tryptophan residue that unambiguously identifies the tryptophan residue.
13. If there is a lysine residue present in the peptide, a strong ε-DNS-lysine spot will be seen when the dansyl derivative of the *N*-terminal amino acid is studied. However, when later residues are investigated the ε-DNS-lysine will be dramatically reduced in intensity or absent. This is because the amino groups on the lysine side chains are progressively blocked by reaction with phenylisothiocyanate at each coupling step of the Edman degradation.
14. The reactions of the lysine side chains with phenylisothiocyanate causes some confusion when identifying lysine residues. With a lysine residue as the *N*-terminal residue of the peptide it will be

identified as bis-DNS-lysine. However, lysine residues further down the chain will be identified as the α-DNS-ϵ-phenylthiocarbamyl derivative because of the side chain reaction with phenylisothiocyanate. This derivative runs in the same position as DNS-phenylalanine in the second solvent, but moves to between DNS-leucine and DNS-isoleucine in the third solvent. Care must therefore be taken not to misidentify a lysine residue as a phenylalanine residue.

15. When glutamine is exposed as the new *N*-terminal amino acid during the Edman degradation, this residue will sometimes cyclize to form the pyroglutamyl derivative. This does not have a free amino group, and therefore effectively blocks the Edman degradation. If this happens, a weak DNS-glutamic acid residue is usually seen at this step, and then no other residues are detected on further cycles. There is little one can do to overcome this problem once it has occurred, although an enzyme that cleaves off pyroglutamyl derivatives has been reported (*2*).
16. The main disadvantage of the dansyl-Edman method compared to the direct Edman method is the fact that the dansyl method cannot differentiate acid and amide residues. Sequences determined by the dansyl-Edman method therefore usually include residues identified as Asx and Glx. This is most unsatisfactory since it means the residue has not been unambiguously identified, but often the acid or amide nature of an Asx or Glx residue can be deduced from the electrophoretic mobility of the peptide (*3*).
17. Having identified any given residue it can prove particularly useful to carry out the procedure referred to as "double dansylation" on this sample (*see* Chapter 23). This double-dansylated sample will identify the amino acids remaining beyond this residue. Double dansylation at each step should reveal a progressive decrease in the residues remaining in the peptide, and give an excellent indication of the amount of residues remaining to be sequenced at any given cycle.

References

1. Hartley, B. S. (1970) Strategy and tactics in protein chemistry. *Biochem. J.* **119,** 805–822.
2. Uliana, J. A., and Doolittle, R. F. (1969) *Arch. Biochem. Biophys.* **131,** 561.
3. Offord, R. E. (1966) Electrophoretic mobilities of peptides on paper and their use in the determination of amide groups. *Nature* **211,** 591–593.

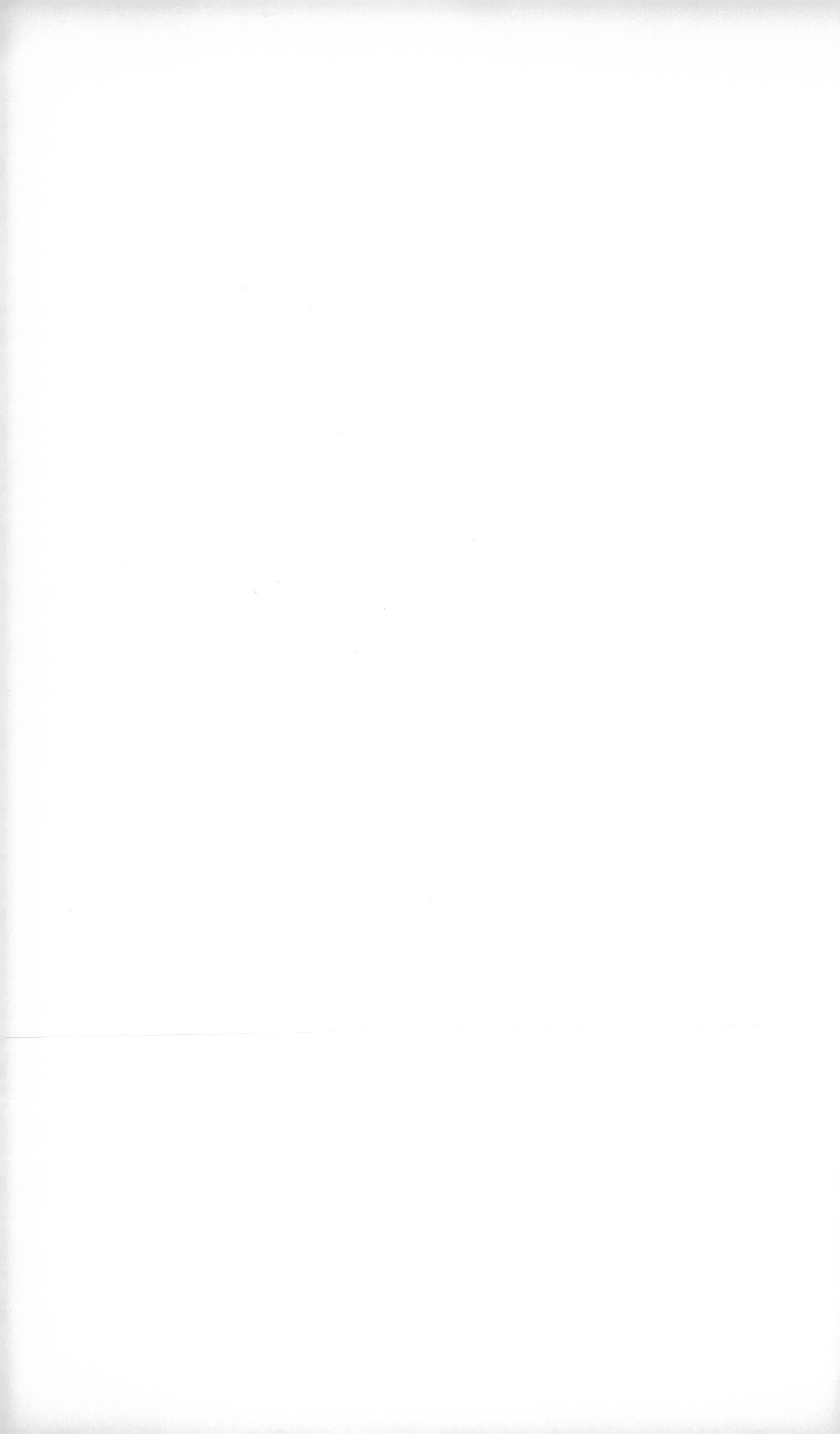

Chapter 25

Microsequencing of Peptides and Proteins with 4-*N*,*N*-Dimethylaminoazobenzene-4′-isothiocyanate

Brigitte Wittmann-Liebold and Makoto Kimura

Max-Planck-Institut für Molekulare Genetik, Abteilung Wittmann, D-1000 Berlin 33 (Dahlem), West Germany

Introduction

Manual methods for the stepwise *N*-terminal degradation of polypeptides have been widely applied in protein chemistry. The technique most commonly used up to now, has been the Edman degradation, carried out either manually (*1*) or automatically (*2–4*). Although nowadays the reaction can be performed with high efficiency and automatically in a sequencer, the manual methods are still of value. The reasons for this are: (i) The manual methods can be easily applied in any laboratory, even by an inexperienced researcher, with a minimum of equipment at low cost; (ii) It is possible to simultaneously screen many peptides for purity, and to gain information in a reasonable time about the *N*-terminal sequences of sets of peptides generated by proteolytic or chemical cleavage; (iii) The selection of fragments of high purity for the sequencer or those peptides that need to be sequenced in a machine because of length, hydrophobicity, or difficult sequence stretches is made easier.

For many years, the manual Edman degradation has been performed either directly (see Chapter 26) or by the dansyl-Edman technique (see Chapter 24). The latter technique has been applied to thousands of peptides and has proved especially valuable when proteins available in submicromol amounts must be sequenced. However, this method has several drawbacks, such as the unavoidable deamidation of glutamine and asparagine during acid hydrolysis of the dansyl-peptide (see Chapter 24 for a detailed discussion of this method).

In 1976, Chang et al. proposed a new chromophoric reagent for protein sequence analysis, 4-*N*,*N*-dimethylaminoazobenzene-4′-isothiocyanate (DABITC), which allows the identification of the released amino acids as red-colored derivatives (*5,6*). Several other modified Edman-type reagents have been proposed for sequential degradation of peptides; however, it was shown that isothiocyanate derivatives with bulky side chains couple less efficiently than PITC itself. Also, in the case of the DABITC reagent, coupling efficiency was low, about 60%, for a reasonable reaction time and temperature, e.g., 40 min and 50°C. Therefore it was necessary to complete the coupling with PITC prior to the cleavage reaction (*7*) in order to avoid severe carryover during degradation. This DABITC/PITC double coupling technique was adequate for practical sequencing in the low nanomol range. In the past few years (*8,9*), these improvements have enabled the complete analysis of the primary structures of all 53 proteins derived from the *Escherichia coli* ribosome. In this chapter we describe the DABITC/PITC coupling method as it is currently being used for microsequencing. Further, we describe the use of this coupling method in connection with manually performed solid-phase degradations.

Materials

Since the use of high purity reagents is necessary for optimal results when sequencing proteins and peptides, purification procedures for most of the reagents used are included in this section.

1. The degradations are carried out in thick-walled glass-stoppered centrifugation tubes (borosilicate glass) and the peptides to be degraded remain in these tubes during the degradation process. The tubes are narrowed at the bottom and have the following sizes: for 5–10 nmol of peptide: length, 45 mm and i.d., 0.8 mm; for 300 pmol to 2 nmol: 35 mm × 0.3 mm. In the latter case, micro-injection tubes, suitable for 5–100 μL volumes, and fitted with silicon stoppers are used. All tubes must fit snugly into thermostatically heated blocks maintained at 55 °C.
2. For all centrifugation steps use a bench centifuge (e.g., Labofuge I supplied by Heraeus/Christ) equipped with adaptors capable of holding the stoppered tubes.
3. Two vacuum pumps that provide a vacuum of 2 Pa measured at the pump inlet. One pump is used for the removal of alkaline solvents after the coupling and the other for the removal of the acid after cleavage. Each vacuum line has a glass distributor for at least six desiccators; the acid vapors are trapped by KOH, and both pumps are protected by a cryostat placed in the vacuum lines.
4. For all experiments, nitrogen having 99.999 vol %, and containing less than 2 vpm = 2×10^{-4} vol % oxygen should be used.

5. Polyamide thin-layer sheets (DC-Mikropolyamide Foils, F1700, 15 × 15 cm, purchased from Schleicher & Schüll) or silica gel thin-layer sheets (HPTLC Kieselgel, 60 F254, 10 × 10 cm, from Merck) are used for the identification of the DABTH-amino acid derivatives. A paper cutter is used to cut the 5 × 5 cm, or 2.5 × 2.5 cm-sized sheets.
6. *p*-Phenylisothiocyanate (PITC) of the fourth highest purity grade is redistilled under reduced pressure (oil pump) and nitrogen, while keeping the destillate at 4°C. Small amounts for the manual degradations are kept in sealed glass ampules under nitrogen at −20°C; larger amounts are stored as stock in 3–4 mL portions also in ampules, under identical conditions.
7. 4-*N*,*N*-dimethylaminoazobenzene-4′-isothiocyanate (DABITC) is recrystallized from acetone and 1 g is then dissolved in 70 mL of boiling acetone, filtered through a paper filter, and allowed to cool slowly. This procedure yields about 0.7 g of brown flakes with a mp of 169–170°C. The reagent is kept as stock in solid form under nitrogen. For manual sequencing, it is dissolved in redistilled acetone and portions of 0.6 mg are transferred to 1-mL plastic tubes, dried *in vacuo*, and stored at −20°C.
8. *p*-Phenylene diisothiocyanate (DITC) is recrystallized from acetone. A 2 g quantity of DITC is dissolved in 10 mL of boiling acetone and allowed to cool slowly. This yields 1.55 g of needles, mp 129–131°C, that give a clear, colorless solution in DMF.
9. 1-Ethyl-3(3-dimethylaminopropyl)-carbodiimide/HCl (EDC), mp 112–115°C, is used without further purification.
10. CPG-10, controlled pore glass (Serva) 40–80 μm diameter (mesh-size 200–400) with a nominal pore diameter of 75 Å.
11. 3-Aminopropyl triethoxysilane (APTS) (Pierce) is used without further purification.
12. Aminopropyl glass beads (APG): 4 g of CPG beads are degassed *in vacuo* for 2 h at 180°C (water pump), then cooled *in vacuo*. A 30 mL quantity of dry toluene and 3 mL of APTS are added. After degassing, the reaction is performed at 75°C for 24 h in a stoppered flask un-

der nitrogen, with gentle stirring. The resin is filtered on a sintered glass filter and washed alternately with toluene, acetone, and methanol (2 × 50 mL, each). The APG is dried *in vacuo* over P_2O_5 at room temperature and stored at 4°C under nitrogen.

13. *p*-Phenylene diisothiocyanate activated glass (DITC-glass): 1 g of DITC is dissolved in 13 mL of DMF, and 2 g of APG are added in portions, with gentle stirring over a 1-h period. The mixture is kept at room temperature for 2 h and the beads are washed with DMF and methanol (2 × 10 mL, each) on a sintered glass filter. The resin is dried *in vacuo*. Storage is as described for APG.
14. Acetone, pro analysis grade (Merck) A 2.5 L quantity of acetone is passed through a column (id of 3.5 cm) of silica (Silica 100–200 active; filled to 8 cm height) and Al_2O_3 (aluminia, grade neutral, activity I, height 10 cm), and then redistilled over charcoal (pro analysis grade) in a 30 cm glass column filled with glass rings, bp 56°C. The solvent is kept over molecular sieve type-3 Å pellets, 2 mm (Merck).
15. Acetic acid, pro analysis grade (Merck) is used without further purification.
16. *n*-Butyl acetate, pro analysis grade (Merck) is used without further purification.
17. Chloroform, pro analysis grade (Merck), is used without further purification.
18. Dimethylformamide (DMF), pro analysis grade (Merck), is purified as follows: Add to 263 mL of DMF, 34 mL of benzene and 12 mL of H_2O. Redistill and discard the first 60 mL of distillation product, up to 150°C. The product above this temperature is collected and redistilled over P_2O_5 under reduced pressure (water pump). DMF is stored over molecular sieve pellets (see acetone) and under nitrogen.
19. Ethanolamine (EA), synthesis grade (Merck), is redistilled. Its bp at 15 mm Hg is 77°C. Store in ampules under nitrogen at −20°C.
20. Ethyl acetate, pro analysis grade (Merck), is purified and stored as given for acetone (bp 77°C).
21. Formic acid, pro analysis grade (Merck), is used without further purification.

22. *n*-Heptane, Uvasol grade (Merck), is used without further purification.
23. *n*-Hexane, pro analysis grade (Merck), is used without further purification.
24. Methanol, Uvasol grade (Merck) is purified and stored as given for acetone (bp 64°C).
25. *N*-methyl morpholine (NMM), synthesis grade (Merck), is distilled over ninhydrin and redistilled over a 30 cm column filled with glass rings. It is stored in ampules under nitrogen at −20°C.
26. Pyridine (PYR), pro analysis grade (Merck), is thrice distilled, once over KOH, then over ninhydrin, and finally over KOH (bp 114–116°C). It is stored in 100-mL glass stoppered flasks (sealed with parafilm) under nitrogen at −20°C.
27. Toluene, pro analysis grade (Merck), is purified as described for acetone, and stored over molecular sieve.
28. Triethylamine (TEA), synthesis grade from Merck, is distilled from phthalic anhydride (bp 89°C). It is stored in ampules under nitrogen at −20°C.
29. Trifluoroacetic acid (TFA), purum grade (Fluka), is distilled over BaO, redistilled from $CaSO_4 \times 0.5 H_2O$ (dried at 500°C immediately before use) over a 30 cm column filled with glass rings (bp 72–73°C).
30. Attachment buffer 1: *N*-methyl morpholine/TFA buffer, pH 9.5: 5 mL of NMM and 5 mL of H_2O are mixed under nitrogen (resultant pH 11.4). About four drops of TFA are added to adjust pH to 9.5. Use freshly prepared buffer and store it under nitrogen.
31. Attachment buffer 2: pyridine/HCl buffer, pH 5.0: 8 mL of pyridine is added to 80 mL of H_2O, they are mixed under nitrogen, and then 5.3 mL of 32% HCl and H_2O is used to adjust to a final volume of 100 mL at pH 5.0. Keep the buffer under nitrogen.
32. Attachment buffer 3: anhydrous buffer, DMF/TEA: 2 mL volume of triethylamine is added to 200 mL of DMF. Alternatively, attachment of peptide under anhydrous conditions is carried out in DMF alone.
33. Attachment buffer 4: bicarbonate buffer, pH 9.0: 0.2*M* $NaHCO_3$ is adjusted to pH 9.0 with 1*M* NaOH.
34. High performance liquid chromatography (HPLC) of the DABTH–amino acid derivatives is performed on a

250 × 4 mm column filled with Shandon MOS-Hypersil, 5 μm, at 45°C (*see* Fig. 2.)

Methods

Manual DABITC/PITC Double Coupling Method

The following procedure is for peptide amounts of 2–5 nmol. With amounts below 2 nmol it is recommended that volumes about one-quarter of those given here be used and that the reaction be performed in restricted tubes (*see* Materials).

1. To the dried peptide add 80 μL of 50% pyridine in water, 40 μL of DABITC solution (2.3 mg/mL in pyridine), flush with nitrogen, and incubate at 55°C for 30 min.
2. Add 5 μL of PITC and incubate at 55°C for a further 30 min. This addition is necessary to avoid any carryover from one degradation step to the next. However, if the *N*-terminal residue of a peptide alone is wanted (end-group determination), this second coupling can be eliminated (*see* ref. 22).
3. This solution is then extracted with 4 × 400 μL of *n*-heptane/ethyl acetate (2:1, v/v) under nitrogen with stirring. Centrifuge and remove the upper organic layer with care. Great care should be taken to avoid withdrawal of the interphase layer, which often carries peptide precipitate when removing the upper phase. This is the reason why we prefer to make four extractions and to leave about 50 μL of the top layer at each centrifugation step.
4. Following extraction, dry the lower aqueous phase *in vacuo* (~30 min). The complete dryness of the lower phase is crucial since any residual reagent or pyridine will lead to progressive salt accumulation at the cleavage step. An efficient vacuum must therefore be maintained to guarantee short, but complete drying stages

This is especially important if longer polypeptides are being degraded.

5. To the dried sample add 50 μL of anhydrous TFA, flush with nitrogen, and incubate at 55°C for 10 min. TFA is then removed *in vacuo* (~10 min). This completes the cleavage reaction. Repetitions of the cleavage (2–4 times) are recommended in the case of hydrophobic peptide bonds, mainly, if Pro-Pro, Pro-Ile, Pro-Val, Pro-Phe, Ile-Ile, Val-Val, Ile-Val, and Val-Ile are involved.
6. To extract the thiazolinone, add 30 μL of water, flush with nitrogen, add 50 μL of *n*-butyl acetate. Vortex and centrifuge. Remove the upper organic layer, repeat the extraction once, combine the extracts, and dry *in vacuo*. The aqueous phase is also dried *in vacuo* and is subjected to further cycles.
7. Conversion of the dried thiazolinone to the respective DABTH-amino acid derivative is carried out by adding 40 μL of 50% TFA, flushing with nitrogen, and incubating at 55°C for 20 min. The sample is then dried *in vacuo*, redissolved in 2-5 μL of ethanol and aliquots taken for analysis (*see* the section on Identification of the DABTH–Amino Acid Derivatives).

Attachment Procedures to Glass Supports

Attachment of Lysine Peptides via Epsilon-Amino Groups to DITC–Glass

1. Transfer 5–10 nmol of salt-free peptide to a solid-phase tube and dry *in vacuo*.
2. Redissolve the sample in 400 μL of 0.2*M* $NaHCO_3$/NaOH buffer, pH 9.0 (alternatively use attachment buffer 1 or 2). Check the solubility of the peptide by withdrawal of an aliquot for picomole amino acid analysis.
3. Add 10 mg of DITC–glass (prewashed in methanol), flush with nitrogen, degas *in vacuo*, and stir at 50°C for 60 min.

4. To saturate remaining sites, add 20 μL of ethanolamine, flush with nitrogen, and incubate at 50°C for 15 min.
5. Wash the beads twice with water (2 mL) and methanol (2 mL), and then dry *in vacuo* (but *see* note at end of the section on attachment via a Homoserine Lactone Residue).

Attachment of Peptides via C-Terminal Carboxyl Groups

1. Transfer 5–10 nmol of peptide to a solid-phase tube and dry *in vacuo* over P_2O_5.
2. Add 100 μL of anhydrous TFA, flush with nitrogen, and incubate for 15 min at room temperature.
3. Dry the sample *in vacuo* (20–30 min), add 2 mg of EDC in 200 μL of DMF, 10 mg of APG (pre-washed in methanol and DMF), flush with nitrogen, degas, and keep at 40°C for 60 min with gentle stirring.
4. Centrifuge and wash the beads with water and methanol (2 × 2 mL).
5. To saturate the remaining sites, add 20 μL of PITC in 100 μL of DMF and 50 μL of pyridine in 200 μL of DMF. Flush with nitrogen, degas, and incubate at 50°C for 20 min.
6. Centrifuge, wash the beads with DMF and methanol (2 × 2 mL), and dry under vacuum (but *see also* note at end of the section on Attachment via a Homoserine Lactone Residue).

Attachment via a Homoserine Lactone Residue

1. Transfer 5–10 nmol of homoserine-containing peptide to a solid-phase tube and dry *in vacuo* over P_2O_5.
2. Add 300 μL of anhydrous TFA, flush with nitrogen, and incubate at room temperature for 1 h. This will convert all homoserine residues to the lactone. Dry the sample *in vacuo*.
3. To achieve attachment, dissolve the peptide lactone in 300 μL of DMF, add 10 mg of APG (pre-washed in DMF), and 50 μL of TEA, flush with nitrogen, degas *in vacuo,* then incubate at 45°C for 2 h. Centrifuge and dry the sample *in vacuo*.

The final drying of the glass beads after attachment may be omitted if the covalently bound peptide is subjected to degradation immediately. Saturation of the residual amino groups of the APG after peptide attachment should be made with PITC. This can be done with the first coupling of the first degradation cycle if DABITC is replaced by PITC (therefore, no DABTH–amino acid can be obtained for the first residue of the peptide).

The Manual Solid-Phase DABITC/PITC Sequencing Method

The method described here is an improved version of one described previously (*10,11*). The reaction is carried out in larger Edman Tubes (1 × 7 cm) with gentle stirring by means of a stirring bar. A magnetic stirrer is placed below the heating block. Special care has to be taken to prevent physical losses of glass-attached peptide during vacuum drying. This can be a problem when removing the TFA after cleavage. 'Bumping out' of the glass beads *in vacuo* is prevented by means of a glass adaptor fitted with a G-2 sinter glass filter that replaces the stopper during the drying stages and restricts the vacuum. We further recommend that drying should be performed as follows: first, dry for 5 min in a Speed Vac Concentrator, then for 20 min in a desiccator.

1. To the glass-attached peptide add 400 μL of 50% pyridine and 200 μL of DABITC solution (2.3 mg/mL in pyridine). Flush with nitrogen and incubate at 55°C for 30 min.
2. Add 20 μL of PITC, flush with nitrogen, and incubate at 55°C for a further 30 min. (At the first cycle, omit DABITC solution, replace with 50 μL of PITC). This completes the coupling reaction.
3. Centrifuge and remove the supernatant. Wash the beads twice with 500 μL of pyridine and then twice with 500 μL of methanol. Dry *in vacuo* (see above).
4. Add 200 μL of anhydrous TFA, flush with nitrogen, and incubate at 55°C for 8 min. Dry *in vacuo*. This completes the cleavage reaction.

5. The thiazolinone is extracted with 400 μL of methanol, and then 200 μL of methanol. The residual glass-attached peptide is dried *in vacuo* and then subjected to further cycles.
6. The combined methanol extracts are dried *in vacuo,* and the thiazolinone is converted by the addition of 80 μL of 50% TFA. The tube is flushed with nitrogen and incubated at 55°C for 20 min. The converted sample is then dried *in vacuo* and identified as described below.

Identification of the DABTH–Amino Acid Derivatives

Two-Dimensional Thin-Layer Chromatography (23)

1. Standard markers for chromatography are prepared as follows: add 500 μL of pyridine to a test tube and 30 μL each of diethylamine and ethanolamine. Add 250 μL of DABITC solution (2.3 mg/mL in pyridine) and incubate at 55°C for 1 h. Dry *in vacuo,* dissolve in 1 mL of ethanol, and apply a small spot together with each sample.
2. Dissolve the DABTH extract in 2–5 μL of ethanol (depending on the initial amount of peptide) and load 0.5 μL aliquots onto polyamide sheets (2.5 × 2.5 cm) together with a standard marker sample. It may be necessary to spot more with increasing number of degradation cycles because of extraction losses of peptide.
3. The chromatogram is then developed in two dimensions (1 min each). The solvents are:

   1st dimension: 33% acetic acid in water
   2nd dimension: toluene : *n*-hexane : acetic acid; 2:1:1, v/v/v

4. All DABTH-derivatives except DABTH-Ile/Leu can be separated with these solvent mixtures as illustrated in Fig. 1. They are visible as red spots after exposure to acid vapors (hold over a flask containing 12*M* HCl) whereas byproducts of the degradation are not visible or form blue-colored spots. The positions of the released DABTH-derivatives are correlated to the posi-

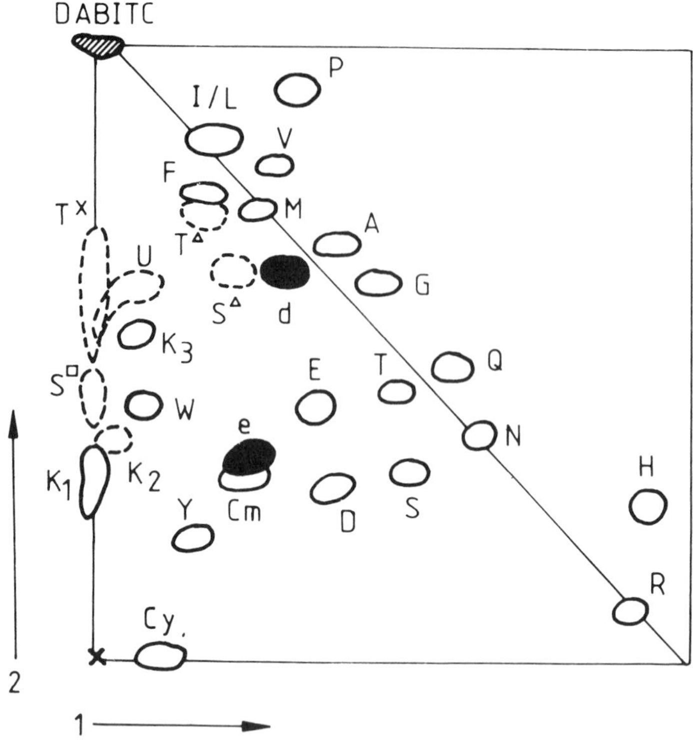

Fig. 1. Two-dimensional thin-layer chromatography of DABTH–amino acid derivatives. About 20–50 pmol of standard DABTH-amino acid derivatives are applied to 2.5 × 2.5 cm polyamide sheets (Schleicher and Schüll). Details of the chromatography systems and the marked spots are given in the text. The DABTH–amino acid derivatives are denoted by the single-letter code; T^{Δ}, dehydrated DABTH-threonine; T^{x}, product formed after β-elimination of DABTH-Thr; $S^{\triangle}$ and $S^{\square}$, corresponding products formed by DABTH-Ser; U, thiourea derivative; K_1, α-DABTH–epsilon-DABTC–Lys; K_2, α-PTC–epsilon-DABTC–Lys; K_3, α-DABTH–epsilon-PTC–Lys; spots "e" and "d" are blue-colored reference markers.

tions of two markers, DABITC-reacted diethylamine (spot "d" of Fig. 1) and ethanolamine (spot "e" of Fig. 1).

5. Thin-layer separation of DABTH-Ile and DABTH-Leu can be performed by one-dimensional chromatography on silica sheets (HPTLC, see materials section) using

formic acid:water:ethanol; 1:10:9, v/v/v, as the solvent mixture (*12*). As this type of thin-layer needs more material only isoleucine and leucine derivatives released in amounts from more than 5 nmol starting peptide material can be identified by this means. Alternatively, HPLC separations, as listed below, may be employed.

HPLC Separation of DABTH-Amino Acid Derivatives

DABTH-derivatives can be separated by isocratic elution in about 30 min (*13*). The self-packed column contains C_8 reversed-phase material, MOS-Hypersil, pore size 5 μm (Shandon) and has a dimension of 250 × 4.6 mm. The solvent system used is 50% 12 m*M* Na acetate, pH 5.0/50% acetonitrile/0.5% 1,2-dichloroethane. The flow rate is 1.2 mL/min at 45°C. The separation of the DABTH-amino acid derivatives is presented in Fig. 2. DABTH-tryptophan elutes between valine and proline, and the derivative of histidine after alanine. DABTH-arginine is retained on the column under these conditions, but can be determined in a separate injection. It is eluted from the column, separated from all other derivatives, with 20% 12 m*M* Na acetate, pH 5.0/80% acetonitrile/0.5% 1,2-dichloroethane. This buffer may also be used to clean the column.

Notes

Manual Liquid-Phase DABITC/PITC Method

1. The general reaction scheme (for the manual method) is presented in Fig. 3. A characteristic color change occurs between the reagent (purple), the thiocarbamoylpeptide (blue), and the released amino acid derivative (red) because of the differences in the resonance structures of the dimethylaminoazobenzene ring after exposure to the acid.
2. The method described can be applied to peptides and proteins. Care has to be taken to avoid salt contamination, as this prolongs all the drying stages, and can prevent the sample from being dried at all, in a reasonable time. After enzymatic cleavages peptides are

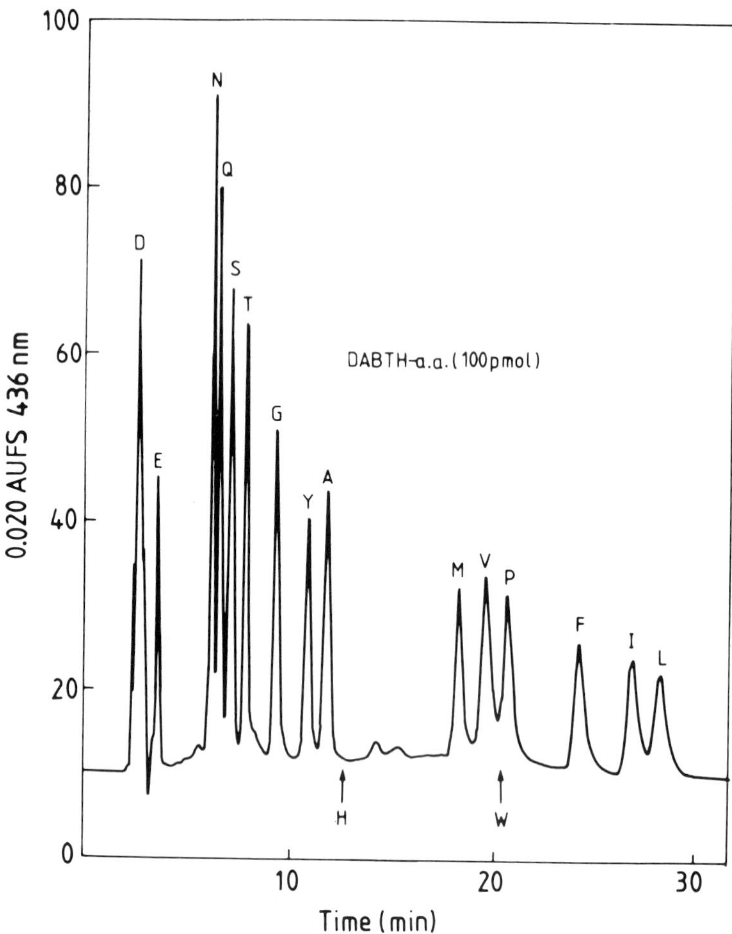

Fig. 2. Isocratic HPLC-separation of standard DABTH-amino acid derivatives. Separation was made on Shandon support, Hypersil MOS (C_8), 5 μm; column size was 250 × 4.6 mm, that of the pre-column (filled with the same support) was 40 × 4.6 mm; flow rate was 1.2 mL/min. Elution at 45°C was made with 50% 12 m*M* Na acetate buffer, pH 5.0/50% acetonitrile/0.5% 1,2-dichloroethane. Injected were 100 pmol of a standard amino acid–DABTH mixture; measurements were made at 436 nm and 0.02 AUFS. All derivatives can be separated under these conditions, see text. However, DABTH-Arg does not elute from the column with this solvent mixture, but migrates separately from all other derivatives by eluting with 20% 12 m*M* Na acetate buffer, pH 5.0/80% acetonitrile/0.5% 1,2-dichloroethane.

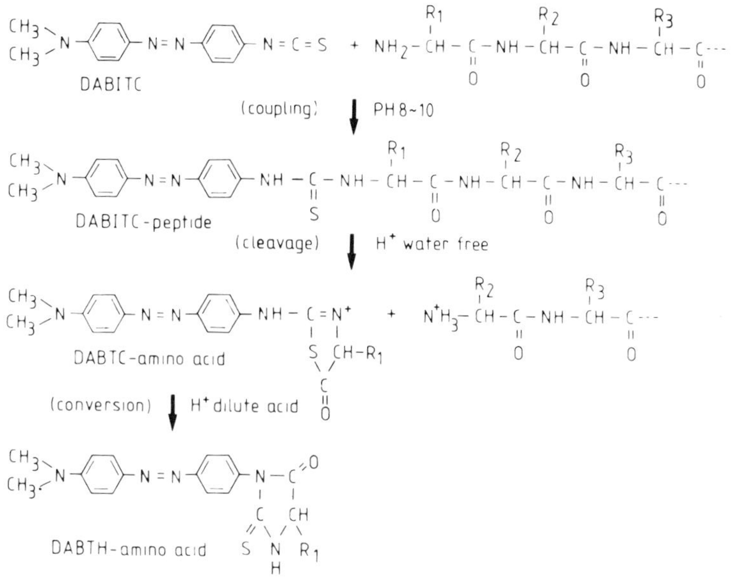

Fig. 3. Reaction scheme of the Edman-type degradation of polypeptides employing 4-*N*,*N*-dir ylaminoazobenzene-4′-isothiocyanate. The degradation consists of three parts: the coupling with the reagent in alkaline medium, the cleavage under acidic, water-free conditions, and the isomerization of the thiazolinone to the hydantoin amino acid derivative.

normally isolated by standard thin-layer techniques (*see* Chapter 4), and their elution is carried out preferably in dilute acids, 0.07% ammonia or 20% pyridine in water (for details, *see* ref. *14*). Peptides deriving from chemical cleavages are desalted on Sephadex columns prior to microsequencing. Alternatively, modern HPLC techniques may be employed for peptide purification (*see* Chapter 5).

Similarly, proteins should be isolated without recourse to ammonium sulfate precipitations, SDS or any method using high amounts of salt. Most suitable are proteins isolated by gel filtration on Sephadex, or by HPLC-methods employing buffers of low salt concentrations (see Chapter 2). Examples where proteins isolated from HPLC columns have been directly used for microsequencing are presented elsewhere (*15,16*).

3. The length of a polypeptide does not limit the application of the method, whereas with the dansyl-Edman technique, difficulties arise with larger sized peptides or proteins. The formation of a red-colored product with DABITC, which is easily visible by eye on thin-layer sheets, down to 20 pmol is a considerable advantage. On the other hand, byproducts of the reaction, which are visible in UV light, are not seen. All polypeptides containing asparagine or glutamine are preferably degraded with this method. Here, the choice of the conversion medium is important: TFA in water (30–50%) produces high yields of the amide derivatives (DABTH-Asn and -Gln), free of the acid derivatives, provided the time for the reaction and the temperature of the block are optimized with standard peptides. Further, arginine and histidine residues are recovered in good yields. On the other hand, serine containing peptides give higher yields of DABTH-Ser if the conversion is performed in 1*M* HCl, or acetic acid, saturated with HCl gas.
4. Serine residues give several spots on the thin-layer chromatogram and lysine yields a number of products because of the use of two coupling reagents, DABITC and PITC, and the presence of α- and epsilon-amino groups during the reaction (*see* Fig. 1). When sequencing less than 1 nmol of peptide or protein, those amino acids that produce more than one spot (lysine, serine, cysteine, and threonine) are difficult to identify or not seen at all. Cysteine can be identified after oxidation or alkylation of the residue in the peptide.
5. Because of the extractions, hydrophobic peptides or those that have the hydrophilic residues in the first part of the sequence are washed out during extraction. In such cases it is better to attach the peptide covalently to a support prior to manually performing the DABITC/PITC degradation.

Manual Solid-Phase DABITC/ PITC Method

6. Peptide attachment to solid supports is essentially as described for the use in solid-phase sequencers (*17*).

In principle, all the supports described for solid-phase sequencing, such as polystyrene or glass, may be used. However, the cleanest results are obtainable for peptides that are linked to glass supports (*18*). Therefore, we have described only these attachment procedures.

7. All types of peptides including the hydrophobic ones are suitable for the solid-phase DABITC method provided they contain a group for the attachment. Best peptide attachment has to be selected according to the solubility and the chemical nature of the peptide; all peptides containing a C-terminal lysine are preferably attached via the epsilon-amino group to DITC-activated glass (*17*) (with yields of 80% or better); all small peptides with free carboxyl-termini but lacking a C-terminal lysine can be attached to APG via carboxyl-activation with EDC (*19,20*), with sufficient yield (approximately 50–80%); peptides containing homoserine are bound to APG (*17*) in yields above 80%. Proteins are best attached via their side chain lysine residues. If several lysines are present, a quantitative reaction of all lysines is not necessary for a tight binding of the protein to the glass. Such bound proteins may be satisfactorily sequenced. It is frequently observed that the sequencing ability of the bound protein decreases with increased reaction of all lysines. Polypeptides eluted from gels, even in the presence of SDS, can be subjected to solid-phase sequencing by the manual DABITC method.
8. All peptides that are attached to the support in good yields can easily be sequenced to their C-terminal end by the manual method. Only in cases of steric hindrance, because of hydrophobic sequences involving proline, isoleucine or valine, is a degradation with repeated cleavages necessary to avoid severe carryover. A peptide or protein may be attached to glass in the presence of SDS (about 0.1%), which is often necessary for hydrophobic peptides. The easiest attachment is under waterfree conditions in DMF, with or without the addition of a base, after pretreatment with TFA (*see* Methods).
9. Limits to the solid-phase method arise from the low solubility of some of the polypeptides during the at-

tachment. Sometimes, insoluble polypeptides behave better with the liquid-phase DABITC technique. Even if they are not soluble in 50% pyridine, they react with the DABITC/PITC reagents to a certain degree, and an efficient degradation can be obtained if the coupling (and extraction thereafter) is repeated several times. Polypeptides that form a separate layer between the phases or at the bottom of the tube can still become degraded.

10. All peptides lacking a lysine (or aminoethylated cysteine) in the C-terminal region, a C-terminal homoserine or a free C-terminal carboxyl-group (carrying a C-terminal amide) cannot be attached to a solid support by the methods described. Covalent binding to a support may then be achieved via tyrosine, tryptophan, or the cysteine-side chain. Carboxyl-attachment is limited by lower yields. Further, peptides with C-terminal glutamine and proline cannot be linked via their C-terminal carboxyl-group to APG. However, the presence of several side chain carboxyl-groups neither disturbs this attachment nor sequencing of the peptide significantly. Under mild attachment conditions (attachment at room temperature for 30 min) at least 30–50% of glutamic acid and aspartic acid residues can be positively identified upon degradation. Attachment via the carboxyl terminus of a protein is also possible, but although good attachment yields have been observed, the sequencing results are poor.
11. The glass beads used for coupling are stable except in the presence of strong bases, especially at high temperature. Therefore, attachment should be carried out under the conditions described in the Methods section, and the coupling buffers should be kept below pH 10. The glass supports decrease in stability in TFA, as they are sensitive to hydrogen fluorides. Therefore, all cleavage times should be kept as short as possible and the acid removed quickly.

Identification of the DABTH-Amino Acid Derivatives

12. Steel holders for drying 10 sheets after chromatography (by means of a cold fan) are helpful. Plexiglass

holders have the advantage that they are acid resistant and may be placed in a desiccator over HC1 vapors for color development of the amino acid derivatives on the sheets. The best solvent chambers for the small sheets are flat-bottomed 50-mL beakers. The rim of the beaker is cut away and the cut surface polished, so that a polished glass plate can be used to close the chamber.

13. The color difference between DABITC, DABTC-peptides, and DABTH-derivatives greatly facilitates the identification of released amino acids. On thin-layers, 10 pmol of the DABTH-amino acid derivatives is the lower detection limit; quantitative determinations by HPLC can be made with at least 20 pmol.
14. Serine and threonine give additional spots, namely the dehydrated derivatives, and an additional blue colored spot, T^x, near thiourea (marked with U); lysine forms as main product α-DABTH–epsilon-PTC–lysine, a red spot which is marked as K3 in Fig. 1; homoserine moves to an almost identical position as DABTH-Thr, but can be differentiated because of the missing blue-colored extra spot described for serine and threonine above. Carboxymethylated cysteine can be identified as a spot that moves into the position of the ethanolamine marker, as shown in Fig. 1.
15. Problems with the identification of the DABTH–amino acid derivatives can be caused by:

    (a) Low purity of the DABITC reagent, which is indicated by the formation of an extra spot in the left upper corner of the chromatogram.
    (b) Inefficient extraction after the coupling leads to a similar spot, which disturbs the migration of the DABTH-derivatives.
    (c) The occurrence of double spots for each of the hydrophobic residues, because of incomplete conversion, or destruction (after too long an exposure to the dilute acid).
    (d) Fast migration of the spots into the front of the second dimension, because the solvent mixture quickly changes its composition (second dimension solvent mixture should be stored at 4°C and used for less than 1 h in the small beaker).
    (e) Low resolution of spots, mainly in the area

DABTH-His/Arg, caused by salt contamination. This can result from increasing salt accumulation during the degradation (incomplete or inefficient drying) or from salt contamination of the original DABTH-amino acid extract. In both cases, it is helpful to dry the DABTH-amino acid extract, to reextract it with water/butyl acetate, and to repeat the identification.

Application to Ribosomal Proteins

16. The DABITC methods described here have been applied to a large number of proteins and peptides derived from different sources over the last few years. Most of the peptides from the *E. coli* ribosomal proteins, whose sequences were finished in 1979–1982, were sequenced using these techniques. Especially difficult sequence areas can be determined in this manner. Further, purity controls of many ribosomal proteins isolated from other organisms were carried out using the manual method, in order to identify which proteins were pure enough to merit their sequencing in a sequencer.

More recently, the DABITC techniques have enabled the complete sequence analysis of about 16 ribosomal proteins isolated from *Bacillus stearothermophilus* in a rather short period of time and using only a few milligrams of material (*21*). The sequence determination of protein S5 from *B. stearothermophilus* may serve as an example (*24*). This protein has been manually sequenced exclusively by the liquid and solid-phase methods. Long sequence stretches of up to 30 residues were sequenced after attachment of the peptides via the C-terminal carboxyl-groups to aminopropyl glass. The combined manual techniques enabled to complete the sequence of this protein in a few months using only 3–4 mg. In summary, all the methods described here are applicable to any type of protein, and the inexpensive instrumentation makes microsequencing possible in almost all laboratories.

Acknowledgment

We are grateful to Mr. Keith Ashman for carefully reading the English version of the manuscript.

References

1. Edman P. and Henschen A. (1975) Sequence Determination, in *Protein Sequence Determination,* Needleman S. B., ed., Springer Verlag, Berlin, pp. 232–279.
2. Edman P. and Begg G. (1967) A protein sequenator. *Eur. J. Biochem.* **1,** 80–91.
3. Laursen R. A. (1971) Solid-phase Edman degradation, an automatic peptide sequencer. *Eur. J. Biochem.* **20,** 89–102.
4. Hewick R., Hunkapiller M., Hood L. E., and Dreyer W. J. (1981) A gas–liquid solid phase peptide and protein sequenator. *J. Biol. Chem.* **256,** 7990–7997.
5. Chang J. Y., Creaser E. H., and Bentley K. W. (1976) 4-*N,N*-Dimethylaminoazobenzene-4′-isothiocyanate, A new chromophoric reagent for protein sequence analysis. *Biochem. J.* **153,** 607–611.
6. Chang J. Y. and Creaser E. H. (1976) A novel manual method for protein sequence analysis. *Biochem. J.* **157,** 77–85.
7. Chang J. Y., Brauer D., and Wittmann-Liebold B. (1978) Micro-sequence analysis of peptides and proteins using 4-*N,N*-dimethylaminoazobenzene-4′-isothiocyanate/phenylisothiocyanate double coupling method, *FEBS Lett.* **93,** 205–214.
8. Wittmann H. G., Littlechild J., and Wittmann-Liebold B. (1980) Components of Bacterial Ribosomes in *Ribosomes,* Chambliss G., Craven G. R., Davies J., Davis K., Kahan L., and Nomura M., eds., University Park Press, Baltimore, Maryland, pp. 51–88.
9. Giri L., Hill W. E., Wittmann H. G., and Wittmann-Liebold B. (1984) Ribosomal proteins: their structure and spatial arrangement in prokaryotic ribosomes *Adv. Protein Chem.* **36,** 2–4648.
10. Chang J. Y. (1979) Manual solid phase sequence analysis of polypeptides using 4-*N,N*-dimethylaminoazobenzene-4′-isoth *BBA* **578,** 188–195.
11. Wittmann-Liebold B. (1981) Micro-sequencing by manual and automated methods as applied to ribosomal proteins, in *Chemical Synthesis and Sequencing of Peptides and Proteins*

Liu T., Schechter A., Heinrikson R., and Condliffe P. eds., Elsevier/North Holland, New York, pp. 75–110.

12. Yang C. Y. (1979) Die Trennung der 4,4-dimethylamino-phenyl-azophenylt hiohydantoin-Derivate des Leucins and Isoleucins über Polyamid-dünnschichtplatten im Picomolbereich. *Hoppe-Seyler's Z. Physiol. Chem.* **360** 1673–1675.
13. Lehmann A. and Wittmann-Liebold B. (1984) Complete separation and quantitative determination of DABTH-amino acid derivatives by isocratic reversed phase high performance liquid chromatography. *FEBS Lett.* **196,** 360–364.
14. Wittmann-Liebold B. and Lehmann A. (1980) New Approaches to Sequencing by Micro- and Automatic Solid Phase Technique, in *Methods in Peptide and Protein Sequence Analysis,* Birr Ch. ed., Elsevier/North Holland, Amsterdam, pp. 49–72.
15. Kamp R. M., Yao Z. J., Bosserhoff A., and Wittmann-Liebold B., (1983) Purification of *Escherichia coli* 30S ribosomal proteins by high performance liquid chromatography. *Hoppe-Seyler's Z. Physiol. Chem.* **364,** 1777–1793.
16. Kamp R. M. and Wittmann-Liebold B. (1984) Purification of *Escherichia coli* 50S ribosomal proteins by high performance liquid chromatography. FEBS Lett. **167,** 59–63.
17. Laursen R. A. (1977) Coupling Techniques in Solid Phase Sequencing, in *Meth. Enzymol.* **47,** Hirs C. H. W., and Timasheff S. N., eds., pp. 277–288.
18. Machleidt, W., Wachter, E., Scheulen, M., and Otto, J. (1973), Coupling proteins to aminopropyl glass. *FEBS Lett.* **37,** 217–220.
19. Wittmann-Liebold B. and Lehmann A. (1975) Comparison of various Techniques Applied to the Amino Acid Sequence Determination of Ribosomal Proteins, in *Solid Phase Methods in Protein Sequence Analysis,* Laursen R. A., ed., Pierce Chem. Corp., Rockford, Ill., pp. 81–90.
20. Salnikow J., Lehmann A. and Wittmann-Liebold B. (1981) Improved automated solid phase micro-sequencing of peptides using DABITC. *Anal. Biochem.* **117,** 433–442.
21. Kimura M. and Chow C. K. (1984) Complete amino acid sequences of ribosomal proteins L17, L27 and S9 from *Bacillus stearothermophilus. Eur. J. Biochem.* **139,** 225–234.
22. Chang J. Y. (1980) Amino-terminal analysis of polypeptide using dimethylaminoazobenzene isothiocyanate. *Anal. Biochem.* **102,** 384–392.
23. Chang J. Y. and Creaser E. H. (1977) Improved chromatographic identification of coloured amino acid thiohydantoins. *J. Chrom.* **132,** 303–307.
24. Kimura, M. (1984) Proteins of the *Bacillus stearothermophilus* Ribosome. The Amino Acid Sequence of Proteins S5 and L30. *J. Biol. Chem.,* **259,** 1051–1055.

Chapter 26

Manual Edman Degradation of Proteins and Peptides

Per Klemm

Department of Microbiology, The Technical University of Denmark, Lyngby, Denmark

Introduction

The Edman or phenylisothiocyanate degradation (*1*) has been employed for the determination of the primary structures of peptides and proteins for approximately three decades. The relative simplicity of the method and its high efficiency in the sequential removal of amino acid residues from the amino terminus of a peptide or protein has resulted in a widespread popularity and usage. In spite of the full automatization of the procedure by Edman and Begg in the nineteen sixties (*2*), the manual version of the sequential amino acid degradation still remains a very realistic and efficient alternative.

The major advantages of manual Edman degradation of proteins or peptides reside in the cheapness and ease with which it can be established in any ordinary laboratory. Indeed in its most simple form all that is needed, apart from the necessary chemicals is a desiccator, a good vacuum pump, and a setup for thin-layer chromatography for identification purposes. Furthermore, since all operations are so easy to perform, little or no previous training is required. The manual method cannot compete with the automated versions either in repetitive yield, i.e., the amount of amino acid derivative recovered per cycle, nor in the number of cycles performed per day (unless you want to work overnight) on a specific peptide or protein. Consequently the size of an amino sequence that can be established from the *N*-terminal region of a peptide or protein falls in the range of 10–30 depending on the amount of starting material and the particular sequence. With automatic equipment this result is normally surpassed by a factor of 2. However, several samples, i.e., 6–12 can adequately be degraded simultaneously with little extra work in the case of the manual procedure—a capability automatic sequencing equipment is not endowed with.

Manual Edman degradation can favorably be applied to the following kind of problems:

1. Characterization of *N*-terminal regions for identification purposes.
2. Elucidation of short *N*-terminal primary structures for use in connection with DNA-sequence information for the establishment of reading frames and starting points.
3. Limited sequencing projects, e.g., a peptide hormone.
4. Sequencing of large number of peptides of up to 30–50 residues in connection with elucidation of complete protein sequences.

The reactions involved in the isothiocyanate degradation are depicted in Fig. 1. The entire cycle can conveniently be divided into the following steps: coupling, wash, cyclization, and extraction (Fig. 2).

1. *The coupling reaction.* In this initial reaction, the free α-amino group of the peptide chain reacts with

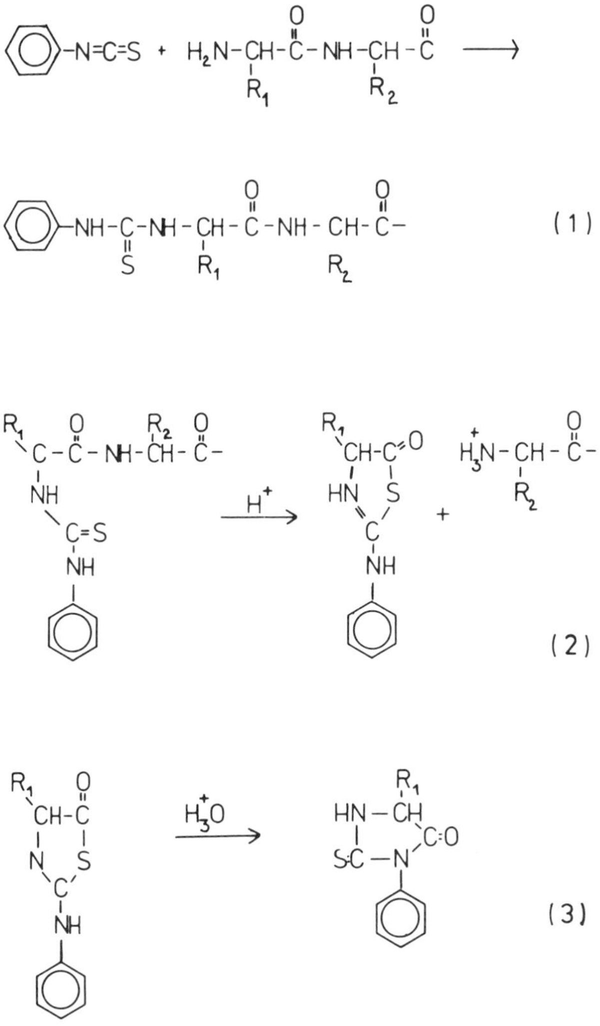

Fig. 1. The Edman degradation reactions and conversion step: (1) the coupling reaction; (2) the cleavage reaction; (3) the conversion step.

phenylisothiocyanate to form a phenylthiocarbamyl derivative (Fig. 1.1). The coupling reaction, which requires a charged amino group ($pK_a \approx 9.5$) is consequently performed in an alkaline milieu, e.g., aqueous pyridine. Furthermore, in order to avoid oxidation, the coupling is profitably carried out in a nitrogen atmos-

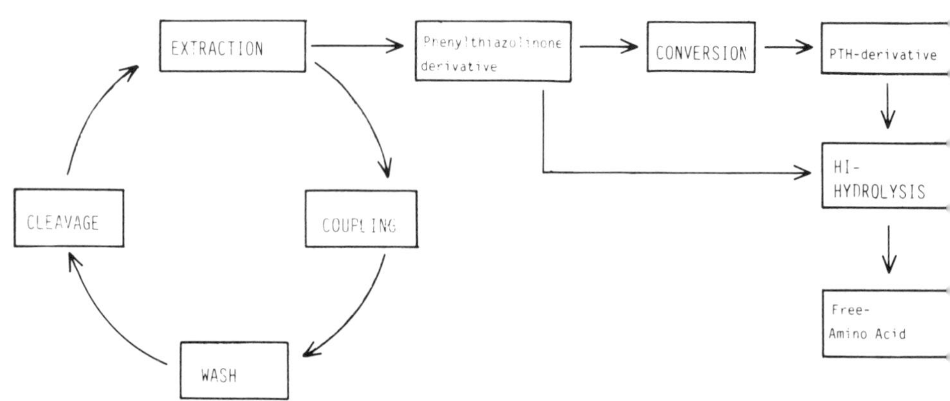

Fig. 2. Diagram showing the steps involved in the manual Edman degradation.

phere. The reaction is generally completed in 10–20 min at 50°C in the case of peptides, but usually takes longer with proteins, in which the α-amino group can be less accessible.

2. *Washing.* In order to remove excess reagent and byproducts from the coupling mixture, the latter is washed with organic solvents, ideally without removing any protein or peptide. In the case of proteins this is normally not a problem. However, with highly hydrophobic peptides the risk of losses resulting from extraction in the organic phase is normally present. If identification by means of the parent amino acid is employed (see later) the washing step can be dispensed of.
3. *Cleavage reaction.* The next reaction (Fig. 1.2) involves cyclization of the phenylthiocarbamyl-peptide and concerted cleavage of the peptide bond. This results in the release of the phenylthiazoline derivative of the first amino acid and demasking of the α-amino group of the second residue. The cleavage reaction requires a strongly acidic milieu in order to proceed. At the same time the presence of water should be avoided to prevent hydrolysis of the peptide chain. The solvent of choice for this reaction is trifluoroacetic acid or other perfluorated carbon acids, which apart from providing an adequate low pH are very good solvents and like-

wise volatile. The cyclization proceeds rapidly at 50°C and is completed within a few minutes, although proline residues are liberated at a slower rate. The conditions used in the cleavage step are unfortunately conductive for dehydration of serine and threonine residues. Special precautions to diminish this have been described extensively elsewhere (3).

4. *Extraction.* After evaporation of the cyclization medium, the residual polypeptide chain is separated from the phenylthiazolinone derivative of the terminal amino acid by liquid partition. This is best performed by extraction of a neutral aqueous solution, e.g., 5% pyridine, that is a good solvent for peptides with a moderate hydrophobic organic solvent like *n*-butyl acetate. The extraction is best carried out on an ice bath in order to diminish side reactions and to provide better separation.

The cycle is then completed and after drying a new series of reactions can be started that bring the Edman chemistry to play on the new *N*-terminal amino acid now constituted by the original second residue.

The phenylthiazolinone derivatives could in principle be used for identification of the respective amino acid residues. However, they are extremely unstable and in practice are therefore not useful for this purpose. Instead, the phenylthiazolinones are converted into the corresponding phenylthiohydantoin (PTH) derivatives, which are quite stable and can be separated and identified by various methods. The conversion reaction (Fig. 1) takes place by heating in dilute acid, e.g., 20% trifluoroacetic acid.

Identification of PTH-derivatives is normally performed by thin-layer chromatography or high performance liquid chromatography (HPLC). Both methods provide excellent resolution and reproducibility.

Thin-layer chromatography of PTH-derivatives is best performed on silica sheets precoated with a fluorescent indicator. Ascending chromatography is carried out with one or two mixtures of organic solvents and takes less than 20 min to do. A reliable system is described in the Method section.

High performance liquid chromatography of PTH-derivatives in reverse phase systems consisting of a hydrocarbon resin eluted with increasing amounts of organic solvent in water provides a fast, reliable and extremely sensitive means for identification, a description of which is outside the scope of this article, but has been treated extensively elsewhere (*4*).

Alternatively, or to supplement the results obtained by analysis of the PTH-derivatives, these can be converted into the corresponding parent amino acids. The regeneration is performed by hydrolysis with hydroiodic acid. If analysis of the parent amino acids is used as a unique means of identification, the thiazolinone derivatives may be hydrolyzed with hydroiodic acid without prior conversion into the PTH-derivatives.

Materials

1. Phenylisothiocyanate is redistilled and stored in ampules of 0.5 mL in the dark at −20°C.
2. Pyridine is likewise redistilled before use and stored at −20°C.
3. Solutions of 2.5% phenylisothiocyanate in 50% pyridine for peptide sequencing are freshly prepared and kept under N_2 at −20°C as working solution for up to 2 wk with no adverse effects.
4. *n*-Butyl acetate-saturated with water, for extraction of phenylthiazolinones, is kept at 0°C.
5. 0.5*M* $NaHCO_3$, adjust to pH 9.8 with 1*M* NaOH.
6. 10% SDS is kept at room temperature.
7. 0.1% ninhydrin, 5% collidine in ethanol for spraying of thin-layer sheets is freshly made prior to use.

Methods

Procedure for Proteins

1. The protein (5–30 nmol) is freeze-dried from a volatile buffer system, e.g., ammonium hydrogen carbonate. The presence of non-amine containing detergents does not interfere with the degradation reactions. A 10 × 75

mm pyrex tube can conveniently be used and all subsequent reactions can be carried out in this.

2. Add 150 μL of 0.5*M* $NaHCO_3$ and 20 μL of 10% sodium dodecyl sulfate. Flush with nitrogen for approx. 20 s, add 10 μL of phenylisothiocyanate, flush again with nitrogen for approx. 20 s, and seal the tube (parafilm or rubber cork). Mix vigorously and place the tube in a water bath at 50°C for 45 min and shake thoroughly at intervals of 5 min.
3. Freeze the tube and add 1 mL of ice-cold acetone (stored in freezer). Shake thoroughly and collect the flocculent precipitate by centrifugation, withdraw and discard the supernatant, and repeat the washing twice. Evaporate the last acetone by flushing with nitrogen while vortexing the tube to distribute the precipitate over the lower part of the tube. Place the tube at an approx. 45° inclination in a desiccator and dry under vacuum for 10 min at 60°.
4. When dry, the sample is ready for the cleavage reaction. Cool to room temperature and add 150 μL of trifluoroacetic acid. Flush with nitrogen for 10 s, seal the tube and incubate for 5 min at 50°C. Evaporate the trifluoroacetic acid by flushing with nitrogen in the hood (use gloves). Dry the precipitate under vacuum in a desiccator for 10 min at 60°C.
5. Place the tube in an ice bath, add 100 μL of ice cold 5% pyridine and extract three times with 300 μL of ice cold *n*-butyl acetate saturated with water (shake and centrifugate). During extraction take great care not to suck up any of the intervening layer separating the two layers where the protein tend to accumulate. Dry the water-phase in the desiccator under vacuum for 20 min. The protein is now ready for the next cycle of Edman degradation.
6. The *n*-butyl acetate containing the thiazolinone derivative is now dried down under vacuum in an desiccator at 60°C (takes approx. 10 min), (Do not leave the tube in the desiccator longer than necessary to avoid side-reactions.) Add 75 μL of 20% trifluoroacetic acid, seal the tube, and leave at 60°C for 15 min (use the desiccator). Dry under vacuum at 60°C for approx. 10 min. The phenylthiohydantoin is now ready for analysis, and will keep well in a freezer for weeks.

Procedure for Peptides

1. Lyophilize the peptide (2–30 nmol) from a volatile buffer in a test tube. Good results have been obtained with drawn-out Pasteur pipets, which provides large inner surface for the amounts of solvent used.
2. For the coupling reaction add 50 μL of 2.5% phenylisothiocyanate in 50% pyridine, flush with nitrogen for 10 s, and seal the tube (small rubber corks are excellent). Incubate for 20 min in a 50°C waterbath.
3. Extract the solution once with 300 μL heptane:ethyl acetate (10:1) and once with heptane:ethyl acetate (1:2). Centrifuge and discard the organic layer. Dry under vacuum at 60°C in a desiccator for 10 min.
4. For the cleavage step add 50 μL of trifluoroacetic acid and leave the tube open in a 45°C desiccator (evacuated for 3 s) for 10 min. The trifluoroacetic acid is removed by applying vacuum to the desiccator for 10 min.
5. To extract the released thiazolinone place the tube in an ice bath, add 50 μL of pyridine and extract three times with 250 μL ice-cold butyl acetate saturated with water. The remaining solution is then dried under vacuum in a 60°C desiccator for 15 min and the peptide is now ready for the next degradation cycle.
6. The thiazolinone present in the *n*-butyl acetate extract is then converted to the PTH derivative as described for proteins (see above).

Identification of PTH-Derivatives

Thin-Layer Chromatography

1. Thin-layer chromatography of PTH-amino derivatives is carried out on silica-gel thin-layer sheets containing fluorescence indicator. The chromatogram is developed in two solvents.
2. Approx. 2 μL aliquots of derivative taken up in methanol are applied by capillary tubes. Ascending chromatography is performed in chloroform/methanol (9:1, v/v) as solvent one and in toluene/methanol (8:1, v/v) as solvent two. Mixtures of standard PTH amino acids are also applied to the chromatogram in lanes adjacent to the unknown sample.

3. The plates are dried in a stream of cold air after each solvent and immediately inspected and photographically recorded under 254 nm light (*see* Fig. 3).
4. Further information can be obtained by spraying the sheets with a solution of ninhydrin/collidine in ethanol, which is then treated in an oven at 110°C for about 5 min before inspection. Using this method, many PTH derivatives develop specific colors, which aids in their identification (*see* Fig. 3). Plates thus stained should be read within 30 min since they tend to turn completely red.

Conversion with Hydriodic Acid

1. Thiazolinone or thiohydantoin derivatives are converted into free amino acids by adding 75 μL of 56% HI to the dried derivative contained in a 7 × 50 mm pyrex tube.

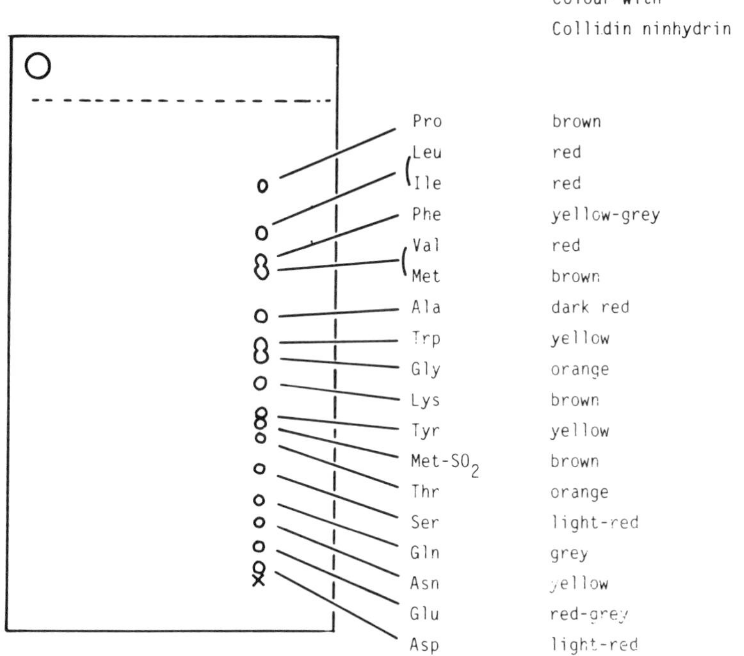

Fig. 3. Separation of PTH-amino acids on silica gel thin layers, showing the colors produced with the ninhydrin/collidine stain.

2. After evacuation the tube is drawn out as an ampule, closed and incubated at 130°C for 10–15 h.
3. The solution is then dried down in a 60°C desiccator under vacuum, and the content submitted to amino acid analysis.

Notes

1. Cycling times are 2 and 1.5 h for the protein and peptide versions, respectively. However, when a large number of samples are processed simultaneously, extra time for handling must be included.
2. The repetitive yields for the presented methods are around 90%, differing slightly from protein to protein.
3. For the best results great care should be taken to use very clean glassware and pure chemicals.
4. Several of the chemicals involved in the Edman chemistry are toxic, notably phenylisothiocyanate and trifluoroacetic acid, and great care should be taken to minimize exposure to these. The extensive use of a hood and gloves is highly recommended.
5. The peptide procedure works well for most peptides up to 50 residues and have even been used for peptides up to 80 residues. Amounts of 2–30 nmol of peptide can be used with the indicated amounts of reagents and solvents. More than 30 residues have been sequenced from a particular peptide with this method.
6. Under the conversion with 20% TFA side chains of Asn- and Gln-derivatives are partially hydrolyzed and the presence of either of these residues will inevitably result in detection of the corresponding acid as well as the amide in roughly equimolar quantities in later analysis. Furthermore, threonine and serine residues, already largely dehydrated during the cleavage reaction, are prone to further degradation in the conversion step. Subsequently, PTH-threonine is normally identified in the form of its dehydrated PTH-derivative. Serine often gives rise to multiple products because of polymerization, which can be seen as multiple tell-tale spots/peaks during later identification.

7. On thin-layer sheets, the PTH-derivatives are visualized under UV-illumination as dark spots, but tend to disappear on prolonged illumination. Consequently, their positions are best registered by photography (green filter) for permanent record. To complement the information provided by the one-dimensional pattern (*see* Fig. 3), the thin-layer sheet can be stained. Several of the PTH-derivatives give quite specific color reactions with the ninhydrin/collidine stain (Fig. 3). As little as 0.1 nmol of a PTH-derivative can be detected by thin-layer chromatography.
8. It is quite normal to load up to a dozen samples on adjacent tracks on a chromatogram, with standard mixtures of PTH amino acids flanking these samples. Do not be alarmed if the chromatogram at first appears complex. Samples will all contain considerable amounts of side-products, particularly phenylthiourea and diphenylthiourea, which will often be present in greater amounts than the PTH-derivative being detected. However, background spots should appear fairly constant in each sample, whereas spots corresponding to the PTH-amino acids are detected by their appearance/disappearance in adjacent samples. Much of the side-product material can in fact be removed by heating the plate at 150°C for 10 min.
9. The hydriodic acid hydrolysis method gives reliable results, but information about amides, methionine, and tryptophan residues is lost and serine and threonine are recovered as alanine and α-amino butyric acid, respectively. Identification of the amino acids can be performed with thin-layer chromatography or preferably with an automatic analyzer.

References

1. Edman, P. (1950) Method for the determination of the amino acid sequence in peptides. *Acta Chem. Scand.* **4,** 283–293.
2. Edman, P., and Begg, G. (1967) A protein sequenator. *Eur. J. Biochem.* **1,** 80–91.

3. Tarr, G. E. (1977) Improved manual sequencing methods. *Meth. Enzymol.* **47,** 335–357.
4. Downing, M. R., and Mann, K. G. (1976) High pressure liquid chromatographic analysis of amino acid phenylthiohydantoins. *Anal. Biochem.* **74,** 298–319.

Chapter 27

Carboxy-Terminal Sequence Determination of Proteins and Peptides with Carboxypeptidase Y

Per Klemm

Department of Microbiology, The Technical University of Denmark, Lyngby, Denmark

Introduction

No useful chemical method (similar to that of the Edman degradation) exists that allows a sequence investigation from the carboxy-terminal end of proteins and peptides. Efforts have therefore been centered around the exploitation of enzymatic methods, notably the use of carboxypeptidases for that purpose. Carboxypeptidases are enzymes that remove amino acids one at a time from the carboxy-terminus of a peptide chain.

Several types of carboxypeptidases have been used in protein chemical investigations, such as pancreatic carboxypeptidases A and B, but their use is limited since they have rather narrow specificities (*1*). However, carboxypeptidase Y (CPY) from bakers yeast has properties that makes it the enzyme of choice for C-terminal investigations. In contrast to most other commercially available carboxypeptidases, CPY has a broad specificity and is able to release all amino acids, including proline, stepwise from the C-terminus of a peptide chain (*2*). Furthermore, since the enzyme retains its activity for extended periods in 6*M* urea or even in 1% sodium dodecyl sulfate solutions, it is excellently suited for the study of proteins that have inaccessible and difficult accessible C-termini under normal assay conditions (*3,4*).

The C-terminal sequence of a given protein or peptide is investigated by studying the kinetics of amino acid release from the peptide chain after addition of CPY. This is done by withdrawing aliquots of the incubation mixture at different time intervals, stopping the reaction (by acidification), and determination of the amounts of free amino acids present in the samples (Fig. 1).

In contrast to the Edman degradation an investigation with a carboxypeptidase seldom results in "linear" sequence data. However, in favorable cases (depending on the protein or peptide), the rate of appearance of the released of amino acids during the digestion will give sufficient evidence to establish the C-terminal sequence. The method presented here has allowed establishment of C-terminal sequences of up to 12 residues flawlessly.

Materials

1. Carboxypeptidase Y (EC 3.4.17.4). The enzyme is stored at −20°C, and solutions of CPY in pyridine acetate buffer are made freshly just before use in each case.
2. 0.1*M* pyridine acetate buffer, pH 5.6, can be stored at 0°C for long periods.
3. Digestion buffer; 0.1*M* pyridine acetate, pH 5.6, containing 1% SDS and being 0.1 m*M* in norleucine.

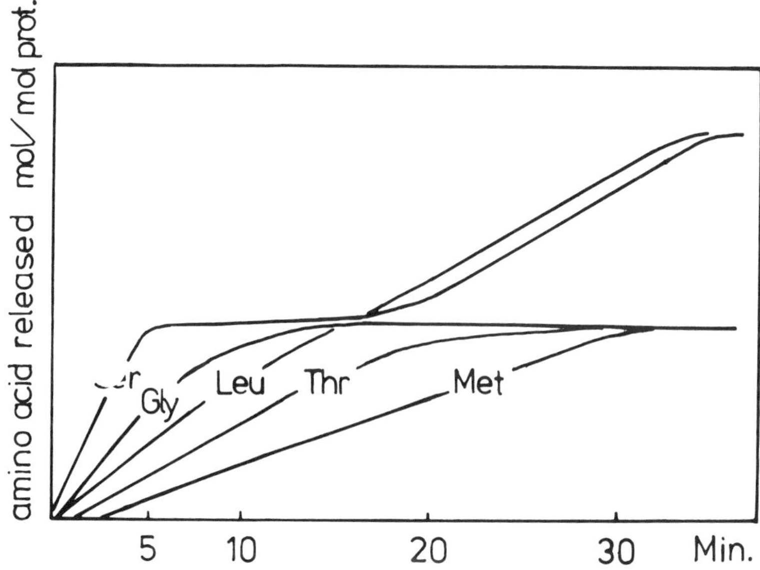

Fig. 1. Rate of release of amino acids from the C-terminal of the CFA/1-protein: Ser - Leu - Val - Met - Thr - Leu - Gly - Ser (4).

Methods

Procedure for Proteins

A prerequisite for the described procedure is access to an automatic amino acid analyzer. The amino acid norleucine (not naturally occurring) is used as internal standard.

1. 20 nmol of protein is lyophilized from a salt-free solution. The protein is dissolved in 200 μL of digestion buffer and the preparation then incubated at 60°C for 20 min in order to denature the protein.
2. After cooling to room temperature a 25 μL aliquot is withdrawn as $T = 0$ background.
3. Incubation takes place at room temperature: 0.2 nmol carboxypeptidase Y (MW of CPY is 61,000) in 5–10 μL

of 0.1*M* pyridine-acetate, pH 5.6, is added, the solution is thoroughly mixed, and 25 μL aliquots are withdrawn at $T = 1, 2, 5, 10, 20, 30$, and 60 min. In order to stop the reaction, 5 μL of glacial acid is immediately added to a sample. Thereafter, it should be frozen and lyophilized. The latter can conveniently be done when all samples have been collected.

4. After freeze drying the samples can be submitted to amino acid analysis, without further preparation. The low amount of sodium dodecyl sulfate applied to the amino acid analyzer does not interfere with the elution profiles, nor with the well-being of the analyzer.

Procedure for Peptides

In the case of C-terminal analysis of peptides, the procedure given for proteins can be applied with the following modifications:

1. Sodium dodecyl sulfate can be omitted from the incubation buffer since the C-terminal of peptides should be readily accessible.
2. The time intervals for sample withdrawal should be shortened since the rate of appearance of free amino acids tends to be faster than in the case of proteins. Use instead $T = 0, \frac{1}{2}, 1, 2, 5, 10, 15$, and 20 min.

Notes

1. Repeated freezing and thawing of solutions of carboxypeptidase Y can lead to denaturing of the enzyme. Likewise, prolonged exposure to room temperature should be avoided as autodigestion of the enzyme can occur. The use of very clean glassware is a prerequisite for good results.

References

1. Ambler, R. P. (1972) Enzymatic hydrolysis with carboxypeptidases. *Meth. Enzymol.* **25,** 143–154.

2. Hayashi, R., Moore, S., and Stein, W. H. (1973) Carboxypeptidase from yeast. *J. Biol. Chem.* **248,** 2296–2302.
3. Gaastra, W., Klemm, P., Walker, J. M., and de Graaf, F. K. (1979) K88 fimbrial proteins: Amino- and carboxyl terminal sequences of intact proteins and cyanogen bromide fragments. *FEMS Microbiol. Lett.* **6,** 15–18.
4. Klemm, P. (1982) Primary structure of the CFA/1 fimbrial protein from human enterotoxigenic Escherichia coli strains. *Eur. J. Biochem.* **124,** 339–348.

Chapter 28

Immunization and Fusion Protocols for Hybridoma Production

J. N. Wood

Department of Experimental Immunobiology, Wellcome Research Laboratories, Beckenham, Kent

Introduction

Kohler and Milsteins' (*1*) technique of monoclonal antibody production is now being exploited in most areas of biology. The essence of the method is to immortalize and then select for clones of plasma cells secreting antibody against a desired antigen. Individual plasma cells secrete antibody of a single antigenic specificity and monoclonal antibodies are thus obtainable from a cell line derived from a single plasma cell. Once established, clonal lines of hybridomas (hybrids composed of plasma cells fused with immortal myeloma cells) provide an infinite supply of antibodies with reproducible properties. Monoclonal antibodies may be produced without extensive purification of the immunogen, and without the tedious cross-adsorption steps often necessary for production of specific antisera. So far, only mouse and rat hybridomas are being produced routinely, while human monoclonal antibodies of potential therapeutic value are technically more difficult to obtain.

Three prerequisites must be satisfied for hybridoma production. It must be possible to induce antibody-secreting plasma cells by immunization in vivo or in vitro. Secondly, a suitable method of immortalization either by fusion with a transformed cell line or direct viral transformation must be available. Thirdly, a quick simple and reliable assay for detecting desired antibodies is necessary for selecting hybridoma clones.

In this chapter the simplest method of immunization and plasma cell immortalization for production of rodent antibodies is outlined. The basic scheme is shown in Fig. 1. Should the antigen of interest be small, say less then 5000 daltons, conjugation of this material to a high molecular weight carrier may be necessary to elicit an adequate immune response. Bovine thyroglobulin (mol. wt. 650,000) is a cheap and effective carrier (4). The method of coupling depends on the availability of reactive groups in the desired immunogen. Here the use of a water-soluble carbodiimide (EDAC) which crosslinks amino and carboxyl groups, and glutaraldehyde, which predominantly couples amino groups, is described. As an example, we have chosen the pentapeptide Met-enkephalin.

To elicit a strong immune response, adjuvants are usually employed during immunization. Freund's complete adjuvant contains heat-killed bacteria to stimulate the animal's immune system, with an oil base in which the immunogen is emulsified to ensure slow release. We describe a protocol for immunizing BALB/c mice.

Where very small amounts of immunogen are available, or human monoclonals are desired, in vitro immunization of cultured plasma cells may be necessary. Some most impressive examples of this technique are now appearing, but it is difficult and beyond the scope of this book (2). While B cell immortalization is usually carried out by polyethylene glycol-induced fusion with a mye-

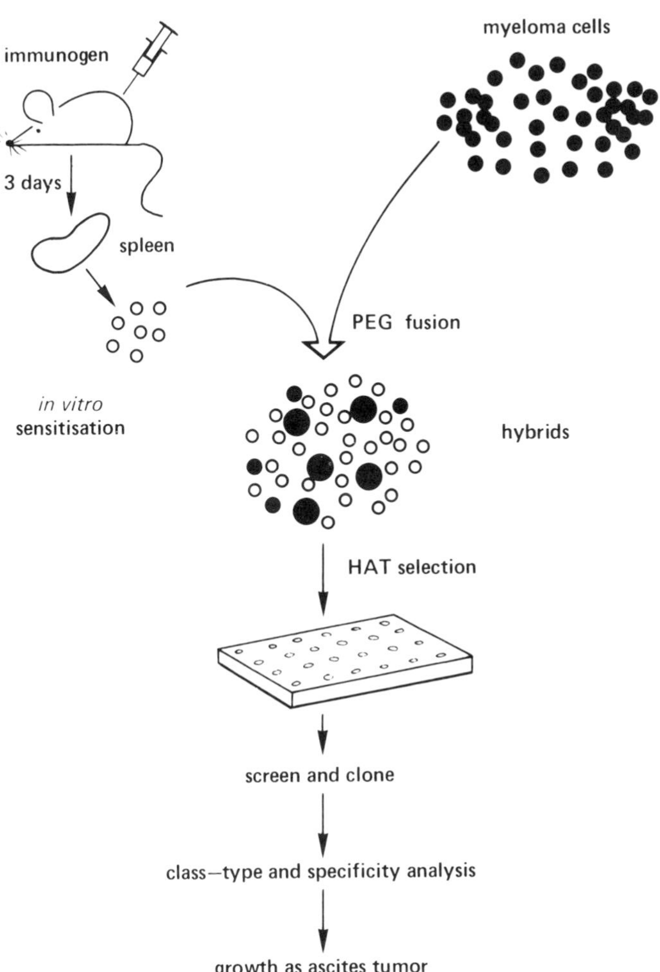

Fig. 1. Diagram showing the steps involved in producing monoclonal antibodies.

loma line, recent developments with high efficiency electric fusion may make this the future method of choice (3).

After polyethylene glycol-induced fusion of immunized splenocytes and myeloma cells, the immortalized hybridomas are outnumbered by residual immortal myeloma cells. In order to kill the myelomas and select uniquely for hybrid growth, the HAT selection system is used. Aminopterin (A) inhibits the *de novo* synthesis of purines by impairing tetrahydrofolate metabolism. In normal cells a second scavenging pathway of purine synthesis, using the enzyme HGPRT (hypoxanthine guanine phosphoribosyl transferase) is present. Hybridoma cells derived from splenocytes can thus still grow in the presence of aminopterin when the scavenging pathway is boosted by exogenous additions of hypoxanthine and thymidine (H and T). However, the myeloma cells used are deficient in the enzyme HGPRT, and they eventually die in the presence of HAT medium. After cell fusion, cells are therefore grown in medium containing HAT for 2 wk, and then transferred to HT medium for a further week (in case residual aminopterin is still blocking the *de novo* synthesis of purines). The $HGPRT^-$ mutant myeloma cell line is obtained by passage in azaguanine (20 μg/mL). Cells with normal levels of HGPRT take up the 8-azaguanine and incorporate it into DNA, leading to cell death.

Simple procedures for the conjugation of haptens, production of hybridomas, their cloning, and storage are presented below.

Materials

Conjugation and Immunization

1. Thyroglobulin (bovine). Store frozen (−20°C or better −70°C) 10 mg/mL in PBS.
2. Glutaraldehyde: 5% solution in PBS made fresh from 25% stock stored at 4°C in the dark.
3. EDAC [1-ethyl-3-(3-dimethyl aminopropyl) carbodiimide hydrochloride].
4. Hydroxylamine: 1*M* solution made fresh in water.
5. PBS (phosphate-buffered saline): KH_2PO_4, 1.5 m*M* (0.2 g/L), Na_2HPO_4, 8.1 m*M* (1.15 g/L), KCl 2.7 m*M* (0.2 g/L), NaCl 140 m*M* (8.0 g/L).

Cell Fusion

1. Polyethylene glycol (1500–4000 form) (PEG) 50% solution in PBS. Store at 4°C in the dark after autoclaving.

2. Growth medium: DMEM or RPMI 1640 supplemented with antibiotics (penicillin 100 U/mL and streptomycin 100 μg/mL), glutamine 2 m*M* and 10% fetal calf serum. Store at 4°C preferably for not more than 1 month.
3. HT medium: Dissolve 272 mg hypoxanthine (H) in 70 mL of 0.1*M* NaOH and make up to 150 mL with water. Dissolve 77.6 mg thymidine (T) in 20 mL H_2O, add to hypoxanthine stock and make up to 200 mL. Filter sterilize, and store at −20°C in aliquots. Dilute in growth medium 100-fold to give a final concentration of H 13.6 μg/mL and T 3.88 μg/mL in HT medium.
4. HAT medium: dissolve 3.82 mg aminopterin in 2 mL NaOH (0.1*N*) make up to 200 mL, filter sterilize, and store in aliquots at −20°C. Dilute 100-fold into HT medium to give a final concentration of 0.19 μg/mL as HAT medium.

Cell Storage and Ascites Production

1. 10 percent analar DMSO (dimethyl sulfoxide) in growth medium, freshly prepared.
2. Pristane, store at room temperature.

Method

Conjugation of Low Molecular Weight Haptens

Carbodiimide Crosslinking

1. Dissolve the antigen in PBS at a molar ratio of 100 to 1 with thyroglobulin (e.g., 1 mg enkephalin, 10 mg thyroglobulin in 1 mL PBS). Add a 200 molar excess of EDAC (0.5 mg) to the mixture, to crosslink amino and carboxyl groups.
2. Mix well and leave at room temperature overnight.
3. Add an equal volume of 1*M* hydroxylamine, and incubate for 4 more hours. This is to reverse a possible modification induced in the tyrosine aromatic ring by the carbodiimide.

4. Dialyze the solution exhaustively against PBS at 4°C, i.e., 100 vol of PBS changed five times.
5. Freeze aliquots of the solution at −20°C, or lyophilize them.

Glutaraldehyde Crosslinking

1. Dissolve the antigen in a 100:1 ratio with thyroglobulin, e.g., 1 mg enkephalin with 10 mg thyroglobulin in PBS.
2. Add 50 μL of 5% glutaraldehyde in PBS and agitate at room temperature for 30 min.
3. Dialyze against PBS at 4°C (five changes of 100 vol of PBS).
4. Store frozen aliquots at −20°C or lyophilize.

In order to estimate the level of binding, it may be useful to add an aliquot of radiolabeled antigen and calculate percentage uptake into nondialyzable material.

Immunization

See also Chapter 32.

1. Emulsify the antigen in Freund's complete adjuvant using an emulsifier, a chopping blade device, ultrasonication, or agitation between two syringes connected by narrow rubber tubing. Use an equal volume of adjuvant to immunogen. A satisfactory emulsion should not separate into phases after prolonged storage at 4°C (weeks).
2. Inject from 1 to 100 μg/animal in about 0.15 mL volume. Intraperitoneal injections are probably the easiest. Immunize several animals if enough material is available.
3. Repeat immunizations with antigen emulsified in incomplete Freund's adjuvant (no bacteria) after about 2 wk.
4. After a further boost without adjuvant, tail bleed the mice for an antibody test, a few hundred microliters should be sufficient.
5. Allow the blood to clot in a microfuge tube containing a glass capillary. After 1 h at room temperature remove the clot with the capillary, and microfuge the sera.

6. Test the sera at various dilutions compared with a normal mouse serum control (*see* Chapters 30 and 31). Ideally a well-immunized mouse should be giving a titer at a dilution of one in a thousand of 50% maximum binding. The mice may be identified by numbering with an ear punch or, less satisfactorily, painting rings on the tails with colored marker pens.
7. When satisfactorily immunized, a final injection without adjuvant should be given 3–4 d before fusion. Intraperitoneal or intravenous injection (tail vein) are satisfactory.

Fusion Protocol

1. Grow HAT-sensitive myeloma cells (commercially available) that do not produce endogenous immunoglobulin (NSO, AG8 653, $SP2_0$ for BALB/c mice, Y123 or YO for Lou rats) in growth medium at a density of about 10^5 cells/mL to give enough cells for a fusion (2×10^6 for a mouse, 1×10^8 for a rat).
2. Kill the immunized animal by cervical dislocation and aseptically remove its spleen 3 or 4 d after a final immunization of antigen without adjuvant.
3. Dissociate the spleen in medium without serum in a laminar flow hood, by homogenizing with a loose potter, teasing with blunt forceps, or passage through a sterile sieve.
4. Wash the splenocytes and the myelomas separately in about 50 mL serum-free growth medium by bench centrifugation.
5. Resuspend both splenocytes and myeloma cells separately in about 25 mL serum-free growth medium and count them with a hemocytometer. A mouse spleen contains between 50 and 200×10^6 lymphocytes, whereas a rat spleen has about 2×10^8 lymphocytes.
6. Mix the cells in a ratio of 1 myeloma to 10 splenocytes for a mouse fusion, or 1 to 2 for a rat fusion. Spin the mixed cells down with a bench centrifuge and aspirate all the medium.
7. Flick the cells around the bottom of the centrifuge tube. Add 1 mL of warm PEG_{1500} solution dropwise over 30 s. Gently agitate the cells for 1 min.

8. Add 20 mL of serum-free medium dropwise over 5 min to the cell pellet, slowly diluting out the PEG.
9. Bench centrifuge out the cells and resuspend in 30–50 mL of warm HAT medium.
10. Plate out the cells at 1–2 × 10^5 cells/well in 96-well microtiter plates. If a complicated screening assay is being used, plate into 24 well-plates. We generally use four plates per BALB/c fusion.
11. After 5 d feed the cells with HAT medium.
12. At 10–14 d (visible colonies) screen the tissue culture supernatants.
13. Expand positive clones in HT medium, from 96 to 24 well-plates, thence to 25 mL flasks. The cells should now be cloned, and some frozen as soon as possible.

Cloning

Positive hybridomas must be grown from a single cell to obtain monoclonal rather than a mixture of antibodies. This cloning process is most easily achieved by growing up hybrids from very low cell densities. The presence of supporting feeder cells is necessary to support growth of the hybrids, which usually die at the low cell densities (~1 cell/well) necessary for cloning. Here we describe a simple and effective cloning method using mouse splenocytes as feeder cells.

1. Tease out splenocytes from stock mouse spleens as described in the fusion protocol. Plate out 4 × 96 well plates/spleen in growth medium containing HT (100 μL/well).
2. The next day, very carefully count a suspension of positive hybridomas. Serially dilute the cells to give concentrations of 0.1, 0.3, 1, 3, 10, and 30 cells/100 μL (about 5 mL of each concentration). Plate out cells at each density in half a 96 well plate (100 μL/well) containing feeders.
3. When visible colonies appear (10 d), screen (Chapters 29,30) and grow several positive wells from the lowest dilutions.

4. Freeze positive cells and repeat the cloning. When a cell line is truely cloned, all wells positive for growth will also be positive by screening.

Freezing Cells

Cells can be stored in liquid nitrogen or, less satisfactorily, in −70°C freezers when suspended at high concentration in 10% DMSO.

1. Make sure the cells are in exponential growth. Bench centrifuge and resuspend in 10% DMSO/growth medium at a density of 0.5×10^6 cells/mL.
2. Freeze down in ampules at 1°C/min by placing the cells in a polystyrene container in a −70°C freezer, or an igloo containing cardice, or the neck of a liquid nitrogen container.
3. When frozen, place in a liquid nitrogen container; be sure the ampules are indelibly marked.
4. Unfreeze cells when required by warming to 37°C quickly. Resuspend then in a large volume of growth medium, bench centrifuge, resuspend, and plate out doubling dilutions in a 24-well plate.

Ascitic Tumor Production

By growing a hybridoma line as an ascitic tumor in an isogenic animal, high concentrations of monoclonal antibody can be rapidly obtained. To diminish the possibility of solid tumor formation, the animals are primed with pristane, a mineral oil.

1. Inject pristane intraperitoneally using a glass syringe 0.2 mL/mouse, 1 mL/rat.
2. At 10 d to 1 month later, inject $2–4 \times 10^6$ cells/animal ip.
3. When the tumor is clearly visible as a large swelling (1–3 wk), tap the ascitic fluid into a heparinized tube using a 21 gage needle, spin out the cells, and freeze the ascitic fluid.
4. Continue to tap the tumors. If the animals are distressed, kill them and aspirate the ascitic fluid with a Pasteur pipet.

Notes

1. It is imperative that animal work comply with legislative regulations and be supervised by a skilled technician. Do not undertake the techniques described here without training by an experienced worker.
2. Hybridomas are easy to make if one is competent at tissue culture. Minor contamination can be knocked out by adding concentrated NaOH to infected wells. Always use the same tissue culture plastic type through the experiments, since changes in plastic type may kill the cells. It is wise to gain experience in a laboratory in which hybridomas are being made.
3. If too many positive wells are obtained hybrids can be frozen down at very low density (*see* ref. *5*).

References

1. Kohler, G., and Milstein, C. (1975) Continuous cultures of fused cells secreting antibody of predefined specificity. *Nature* **256,** 495–497.
2. Reading, C. L. (1982) Theory and methods for immunization in culture and monoclonal antibody production. *J. Immunol. Methods* **53,** 261–291.
3. Zimmerman, U. (1982) Electric field-mediated fusion and related electrical phenomena. *Biochim. Biophys. Acta* **694,** 227–277.
4. Skowsky, W. R., and Fisher, D. A. (1972) The use of thyroglobulin to induce antigenicity to small molecules. *J. Lab. Clin. Med.* **80,** 134–144.
5. Wells, D. E., and Price, P. J. (1983) Simple rapid methods for freezing hybridomas in 96 well microculture plates. *J. Immunol. Methods.* **59,** 49–52.

Chapter 29

Immunofluorescence and Immunoperoxidase Screening of Hybridomas

J. N. Wood

Department of Experimental Immunobiology, Wellcome Research Laboratories, Beckenham, Kent

Introduction

As mentioned in the previous chapter, the provision of a rapid, simple assay for screening monoclonal antibodies is essential for their production. Although solid-phase binding assays are the method of choice (Chapter 30) a number of applications require visualization of antibody binding to cultured cells or tissue sections.

To give some examples, if a cell type can be defined by immunodetection of a specific surface antigen, then purification and characterization of this particular cell type may be possible. Such an approach has been successfully used in defining lymphocyte subpopulations, and determining the origin of neuronal cell types in culture. The ability to visualize a viral antigen may elucidate the role of virus infection and expression in cellular transformation. Spatial or temporal variation in an antigen expression during cell growth or differentiation may also provide useful markers to define different sorts of cellular organization, while disease-state-specific antigen expression may enable the production of diagnostic monoclonal antibodies, even though the nature of the antigen is unknown. This chapter is concerned with the simplest method of fixing cells and tissue sections, and the visualization of antibody binding by fluorescent- or peroxidase-linked second antibodies.

Fluorescent antispecies second antibody, produced by coupling a fluorochrome to the affinity-purified antisera, is used to visualize monoclonal antibody binding to cells or tissue sections by means of fluorescence microscopy.

Indirect immunoperoxidase staining gives a permanent record of antibody binding, and may thus be preferable to immunofluorescence screening, which can fade. In essence, the peroxidase attached to the second anti-species antibody is used to produce an insoluble stain at the site of antibody binding. Excellent commercial enzyme-linked anti-species antibodies are now available. The sensitivity of the technique may be much enhanced by the use of Stenberger's peroxidase–antiperoxidase method (*2*), but for routine screening this should be unnecessary.

There are no hard and fast rules for predicting the best fixation method for an antigen. Acid alcohol, paraformaldehyde, and picric acid are commonly used, as are unfixed frozen sections. A very useful method has been described by Lane (*1*), where cultured cells are fixed and dried on tissue culture plates using acetone/methanol, which does not render the plastic opaque. This enables many samples to be screened on a single petri dish.

As well as describing methods of visualizing antibody staining, a protocol for coupling fluorescent dyes to antibodies is described. Using fluorescein or rhodamine isothiocyanate, the dyes may be directly conjugated to antibodies at alkaline pH without any coupling agent. Peroxidase conjugation is covered in the next chapter.

Materials

Fixation of Cultured Cells

1. 90% Methanol/10% acetic acid. Store at −20°C.
2. 50% Acetone/50% methanol. Store at room temperature.
3. 40% Formaldehyde. Store at room temperature. Dilute 1:10 in phosphate-buffered saline (PBS) for working solution (*see* Chapter 28).

Tissue Sections

1. 40% Formaldehyde (see above).
2. 4% Paraformaldehyde. Make fresh by dissolving 8 g paraformaldehyde in 50 mL H_2O at 60°C, adding NaOH (1*N*) dropwise, in a fume hood. Add 50 mL double strength PBS (*see* Chapter 28) when the paraformaldehyde is dissolved.
3. 5% Sucrose in PBS (*see* Chapter 28).
4. Glycerol.

Immunofluorescence

1. Normal goat serum. Store at 4°C with 0.02% sodium azide.
2. PBS (*see* Chapter 28).
3. Goat anti-mouse heavy and light chains (FITC- or TRITC-conjugated) Store frozen aliquots in the dark. Dilute 1:50 in PBS as a working solution and microfuge before use.
4. Mounting medium: 70% glycerol, 0.02% azide, 0.5% NaCl, 0.02% NaOH, 0.4% glycine.

Immunoperoxidase

1. 70% Ethanol
2. 100% Methanol, 0.1% H_2O_2 (make fresh from 30% H_2O_2 stock)

3. Peroxidase-conjugated goat anti-mouse heavy and light chains. Store frozen aliquots, dilute 1:50 in PBS and microfuge before use.
4. 4-Chloro-1-naphthol: 3 mg/mL in methanol. (Store in the dark at 4°C.) Dilute 1:5 v/v in PBS with 0.01% H_2O_2 immediately before use.
5. Mounting medium, *see* above.

Conjugating Fluorochromes

1. Sodium borate (50 m*M*, pH 9.3) containing 0.4*M* NaCl. Store at 4°C.
2. Dimethyl sulfoxide (DMSO).
3. TRITC (Tetramethyl rhodamine isothiocyanate). (Store dry, −20°C in the dark).
4. FITC (fluorescein isothiocyanate). (Store dry, −20°C in the dark).
5. Purified antibody.

Methods

Fixation of Cultured Cells

1. Wash the cell cultures twice in PBS to remove serum.
2. Fix cells in

   (a) 90% methanol–10% acetic acid (−20°C) for 2 min or
   (b) 4% formaldehyde in PBS for 30 min or
   (c) 4% paraformaldehyde in PBS for 30 min.

3. Wash cells in PBS.
4. Proceed directly to antibody staining (2.3) if few samples are being tested. Otherwise continue with steps 5 and 6.
5. Add acetone/methanol (1:1 v/v) for 2 min.
6. Aspirate the acetone/methanol and air-dry.

Tissue Sections

For best results, it may be necessary to perfuse fresh tissue with fixation medium. However, soaking small chunks of tissue in fixative prior to sectioning, or using frozen unfixed tissue sections may be adequate.

1. Immerse fresh tissue in 4% paraformaldehyde overnight at 4°C, using 0.5 cm cube chunks of tissue.
2. Rinse in PBS, and soak in 5% sucrose PBS for several hours.
3. Cut cryostat sections and lift them off the knife onto coverslips smeared with a thin layer of glycerol (a thumbprint).
4. Store sections at −20°C, but try to use them as quickly as possible.

Indirect Immunofluorescence

Cells

1. Spot 5 μL samples of tissue culture fluid in a known order onto a petri dish containing fixed air-dried cells, previously marked with ~0.5 cm square grids.
2. Incubate in a moist box for 1 h at room temperature, being very careful not to allow the samples to run into each other.
3. Wash the plate four times in PBS, with a final longer wash (5 min) in PBS 1% normal goat serum [this is to block any nonspecific second (goat) antibody-binding sites, if you are using rabbit anti-mouse then use 1% normal rabbit serum].
4. Add 1/40 dilution of microfuged goat anti-mouse globulins (FITC or TRITC labeled). Cut down the volume required by covering the petri dish with coverslips when the second antibody has been applied.
5. After 1 h, wash the petri dish five times in PBS, floating away the coverslips. If samples are too near the rim of the dish to examine microscopically, tear off the sides with pliers.
6. Coverslip the samples after addition of mounting medium (seal them with nail vanish), and examine by epifluorescence microscopy.

If you are careful, it may be possible to quickly screen the samples at low power (obviously without using oil or

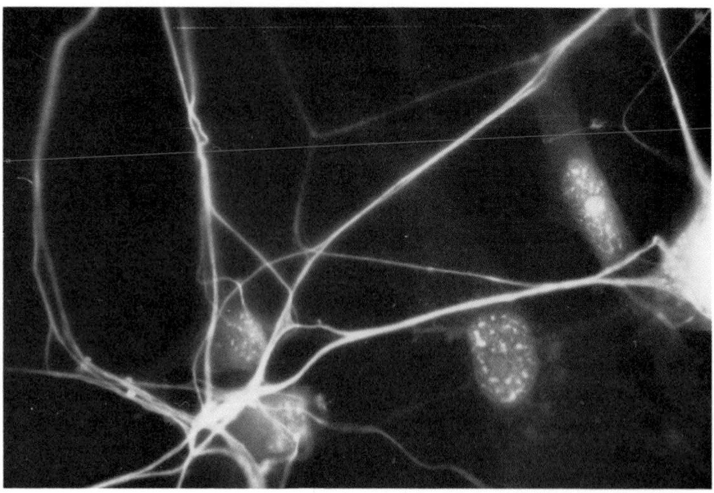

Fig. 1. Indirect immunofluorescence of an anti-neurofilament monoclonal antibody staining rat dorsal root ganglia cultures [×710].

water immersion lenses) directly after washing, and without bothering to mount or coverslip the cells.

Coverslip Mounted Sections or Cells

1. Add 100 μL of tissue culture fluid to the section or fixed cells on the coverslip, and keep them in a wet box for 1 h at room temperature. Number the coverslips with a diamond pen.
2. Using coverslip racks, or Coplin jars, wash the coverslips four times in PBS, and once for 5 min in 1% normal goat serum (see above).
3. Replace the coverslips the right way up (check the numbering) in the wet box, and add 50 μL 1:50 FITC or TRITC goat anti-mouse globulins.
4. After 1 h, wash five times in PBS, mount in mounting medium on slides, seal with nail vanish, and examine.

Immunoperoxidase Screening

1. Using fixed cells in petri dishes, or sections on coverslips, immerse in:

(a) 70% ethanol, 10 min
(b) 100% methanol, 0.1% H_2O_2, 10 min
(c) 70% ethanol, 10 min
This inactivates endogenous peroxidases.

2. Incubate in hybridoma supernatant and wash as described in the immunofluorescence protocol.
3. Add 1/50 dilution of peroxidase anti-species antibody for 1 h at room temperature.
4. Wash four times in PBS.
5. Add 4-chloro-1-naphthol working solution for 10–20 min.
6. Wash in PBS to stop the reaction, and mount in glycerol medium. The reaction should be visible to the naked eye even before microscopic examination.

Conjugation of Fluorochromes to Proteins

With the commercial availability of fluorescent antispecies antisera, indirect immunofluorescence is the usual method of visualizing monoclonal antibody binding to cells or tissue sections. However, direct conjugation of a fluorochrome to a monoclonal antibody may occasionally be useful, for example, in double labeling experiments with monoclonals of similar isotype. Direct conjugation of an enzyme, for example peroxidase, may be useful for a similar reason, and this is covered in Chapter 30.

1. Dissolve purified antibody (2 mg/mL) in 50 m*M* borate, pH 9.3, containing 0.4*M* NaCl.
2. Dissolve 3 mg FITC in 100 mL 50 m*M* borate buffer, pH 9.3. For TRITC, dissolve 3 mg in 0.2 mL DMSO, then make up to 100 mL of 50 m*M* borate, pH 9.3.
3. Dialyze the antibody (1–5 mL) against the dissolved fluorochrome overnight at 4°C in the dark.
4. Dialyze the conjugated antibody against several changes of PBS at 4°C in the dark for 3 more days.
5. Store the antibody in frozen aliquots (−20°C) in the dark.

Notes

1. Cutting sections requires a sharp knife, careful reading of the cryostat instruction manual, and some perseverance. It is easy enough to cut thick (30 μm) sections, which are quite adequate for preliminary screening.
2. To examine surface staining of live cells, it may be necessary to inhibit possible internalization or shedding of the antigen–antibody complex. For this reason, the cells are stained at 4°C in the presence of 0.02% sodium azide. Follow the previous protocol, using chilled reagents, unfixed cells, and 4°C incubations. For suspension cells, incubations and washings can be carried out using small plastic tubes (LP3) and a bench centrifuge. At the end of the staining protocol, small volumes (10–20 μL) can be mounted and screened as usual.
3. One potential problem is the presence of endogeneous peroxidase in some tissues and cell types. For this reason β-D-galactosidase-coupled second antibodies together with a suitable substrate (e.g., *o*-nitrophenyl β-galactopyronoxide) may sometimes be useful. In the protocol described, an inactivation step for endogeneous peroxidase is included although this step may be unnecessary for the cell type that you are investigating.

References

1. Lane, D. P., and Lane, E. B. (1981) A rapid antibody assay system for screening hybridoma cultures, *J. Immunol. Methods* **47,** 303
2. Sternberger, L. A. (1974) *Immunocytochemistry,* Prentice-Hall, Englewood Cliffs, NJ.

Chapter 30

Solid-Phase Screening of Monoclonal Antibodies

J. N. Wood

Department of Experimental Immunobiology, Wellcome Research Laboratories, Beckenham, Kent

Introduction

By far the most convenient and simple assays for screening monoclonal antibody-producing hybridomas are those utilizing solid-phase binding assays, where radiolabelled or enzyme linked antispecies antibodies are used to detect binding of the hybridoma supernatants to insolubilized antigen. Such procedures require an adequate supply of antigen, preferably purified, which can be bound to a solid support, usually through noncovalent binding to PVC microtiter plates or nitrocellulose sheets.

The advantage of such methods lies in their simplicity and speed. Hundreds of samples can be easily screened in a day. After negative clones have been discarded, considerable information about the affinity of the antibody, and species and tissue distribution of the antigen can be rapidly obtained with minor modifications of the assay. This allows a further level of selection to be exercised on the antibody before exhaustive characterization.

In this chapter, simple solid-phase binding assays, iodination techniques, and conjugation of enzymes to antibodies are described.

The two solid-phase assays, which differ in the incubation conditions for attachment of the antigen to plastic (PVC) plates should be compared for their sensitivity with the particular antigen being investigated. Should antigen be in short supply, nitrocellulose has advantages as a solid support, in that very small amounts of antigen may be used, and a fine discrimination between real and background binding can readily be made, either by radioautography or enzyme-linked assays (2).

Although commercial iodinated antispecies antibodies are available, they are expensive, and it may be necessary to prepare one's own. The protocols described below may be used to iodinate either antigens or antibodies via tyrosine aromatic rings to a high specific activity (*see* Chapter 31).

An alternative approach, using enzyme-linked second antibody to visualize antibody binding is equally good. If labelled antigen is available, the assay may be carried out more conveniently by "dotting" the hybridoma supernatants on a grid, and soaking the whole sheet in labelled antigen. Positive wells will be detectable by a dark spot on autoradiography.

Materials

Iodination

Chloramine T

1. Chloramine T, fresh solution: 5 mg/mL in water.
2. Dowex 1 X-8 resin. 2 mL in a plastic syringe freshly equilibrated with PBS (phosphate buffered saline) (Chapter 28) containing 0.1% bovine hemoglobin.
3. Tyrosine in PBS, fresh saturated solution: (1 mL).

4. 25% Trichloracetic acid, store at room temperature (stable).
5. Na ^{125}I high-specific activity (carrier-free) in NaOH. Always use a fume hood when diluting this into PBS for iodinations.

Iodogen

Iodogen (1,3,4,6-chloro-3a,6a-diphenyl-glycoluril).

Make up at 100 μg/mL in dichloromethane and add 25 μL/microfuge tube. Air dry the open microfuge tubes by rotating them briskly, and when dry, store them at −20°C (stable for weeks).

Peroxidase Conjugation

1. Horse radish peroxidase (store dry at −20°C).
2. Immunoglobulin (preferably affinity-purified).
3. Glutaraldehyde: 0.02% made fresh in PBS.
4. Sephadex-G200 column equilibrated with PBS.

Plate Binding Assay

Method A

1. PBS containing 5% bovine hemoglobin and 0.02% sodium azide, store at 4°C.
2. ^{125}I antispecies antibody [e.g., rabbit anti-mouse, preferably F(ab′)2 minimum specific activity 10 μCi/μg]. Stored at 4°C with 0.02% sodium azide.

Method B

1. Adsorption buffer: 50 m*M* bicarbonate, pH 9.6 (1.59 g Na_2CO_3, 2.93 g $NaHCO_3$, 0.2 g NaN_3 in 1 L). Store at 4°C.
2. Wash buffer: Tween 40 (0.05%) in PBS containing 1% hemoglobin and 0.02% NaN_3 (store at 4°C).
3. ^{125}I second antibody, as above.

Enzyme-Linked Assay

1. Peroxidase-conjugated antispecies antibody [preferably F $(ab')_2$] store frozen aliquots after reconstitution, and dilute to 1/60 in PBS and microfuge before use.

2. 4-Chloro-1-naphthol, 3 mg/mL in methanol stored at 4°C in the dark (weeks). Working solution 1:5 v/v in PBS with 0.01% hydrogen peroxide.
3. Hydrogen peroxide 30%. Store at 4°C.

Nitrocellulose Binding Assay

1. 5% Bovine hemoglobin in PBS plus 0.02% sodium azide (4°C).
2. 1:5 dilution in PBS of above.
3. ^{125}I-antispecies antibody [e.g., rabbit anti-mouse, preferably F(ab′)$_2$ minimum specific activity 10 μCi/μg].
   or

3a. Alternatively, enzyme-linked antibody reagents marked under ELISA.

Methods

Iodination

Chloramine T Iodination

This fast oxidation may reduce the antigenicity of some molecules, but is usually satisfactory for iodinating immunoglobulins.

1. In a fume hood, mix the antibody or protein (200 μg in 250 μL PBS) with ^{125}I (sodium salt, 250 μCi in PBS) and 25 μL Chloramine T.
2. After 1 min at room temperature add 50 μL tyrosine in PBS.
3. Chromatograph the solution on a Dowex 1 X-8 column (2 mL in a disposable syringe) equilibrated with PBS containing 0.1% hemoglobin.
4. Collect the column eluate (first 3 mL) and discard the column to which free ^{125}I is bound.
5. Measure incorporation by mixing 10 μL aliquot with 100 μL of 1% hemoglobin in PBS. Add 1 mL 25% TCA, leave for 1 hr at 4°C then microfuge (10,000*g*, 5 min) and measure both the precipitated and soluble counts. The percentage of counts incorporated (precipitate) should be $<$ 80%.

Iodination with Iodogen

This slightly milder process uses iodogen-coated tubes and may be preferable for labelling monoclonal antibodies or labile antigens.

1. Add antibody or antigen (100 μg in 100 μL PBS) to a coated iodogen tube.
2. Add 250μCi ^{125}I (sodium salt diluted in PBS) and mix.
3. After 20 min at room temperature, chromatograph on a Dowex column as described above.

Conjugation of Peroxidase to Immunoglobulin

1. Mix 1 mg of immunoglobulin with 5 mg of horseradish peroxidase in 1 mL PBS.
2. Add 1 mL of glutaraldehyde and mix for 30 min at room temperature.
3. Chromatograph on a column of G200 Sephadex equilibrated with PBS (column size about 25 × 2 cm) and retain the first peak of conjugated antibody. Store frozen aliquots that should be stable indefinitely.

Solid-Phase Binding Assays on PVC Plates

Two protocols are described, one or the other of which has proved satisfactory with protein antigens ranging from small peptides to whole brain homogenates (*1*).

Binding Assay A

1. Pipet 50 μL of antigen (1–100 μg/mL) in PBS into each well of a 96-well PVC plate and incubate overnight at room temperature, to allow antigen to bind to the well surface.
2. Remove the surplus antigen and wash three times with 5% (w/v) hemoglobin in PBS (wash bottle).
3. Fill the wells with Hb solution and incubate for 30 min at room temperature, to block further binding of protein to the surfaces of the wells, then wash twice more with the same solution.

4. Pipet dilutions of antiserum, or samples of hybridoma supernatants into the wells (45 μL/well) and incubate for 1 h at room temperature. Remove the antiserum by aspiration, or more simply by flicking the contents of the plate into a sink, and wash five times with 1% (w/v) Hb in PBS.
5. Pipet the second antiserum (^{125}I-labeled rabbit anti-mouse globulins; 45 μL/well about 20,000 cpm/well) into the wells and incubate 1 h at room temperature. Remove the labeled antiserum and wash six times with 1% Hb in PBS, followed by one wash with PBS. Allow the plate to dry, then cut the wells and count samples with a gamma counter.

Binding Assay B

1. Incubate the antigen (1–100 μg/mL) in adsorption buffer overnight on 96-well plates (50 μL/well) at 4°C.
2. Sink the plate in wash buffer for 2 h to block remaining sites. Rinse the plate briefly in water, and drain. The plates may be stored at −20°C until required, or used immediately.
3. Add dilutions of antisera, or tissue culture supernatants (50 μL/well) overnight at 4°C.
4. Wash four time in wash buffer.
5. Incubate ^{125}I-antispecies antibody (40 μL/well) containing 20,000 cpm for 1 h at room temperature.
6. Wash four times in wash buffer, dry, and count.

Enzyme-Linked Immunoassay (ELISA)

Both of the assays above can be adapted by using peroxidase-conjugated second antibody.

At step 5, substitute 1/60 diluted peroxidase conjugated antibody for a 1 h incubation.

6. Wash four times in PBS.
7. Add working solution of fresh 4-chloro-1-naphthol (50 μL/well).

8. When dark blue spots have appeared, stop the reaction by washing the plate.

Nitrocellulose Assay

1. Draw a 0.5 cm square grid on the nitrocellulose sheet using an ordinary biro, and "dot" the antigen solution centrally with a micropipet (1 μL of 1 mg/mL in PBS). Air dry.
2. Store the sheet desiccated at −20°C until required (months).
3. Number the nitrocellulose squares with a biro, and then soak the grid in 5% Hb for 1 h.
4. Cut out the numbered squares from the grid and incubate them overnight at 4°C in neat tissue culture supernatant or dilutions of antisera (serial tenfold). Use 5 mL bijoux containing about 0.25 mL sample volume.
5. Wash the squares four times in PBS, then add ~20,000 cpm/square of ^{125}I antispecies antibody (10 μCi/μg) in 1% Hb and incubate for one more hour.
6. Wash the squares five times in PBS, dry, and autoradiograph

Notes

1. Neither PVC plates nor nitrocellulose blots should ever be allowed to dry out during assays.
2. Always include a blank plate as a control in these assays, as nonspecific binding is a common problem.
3. Always compare normal serum with immunized mouse serum at various dilutions during an assay to give low and high control points.
4. When starting to use these assays, it may be necessary to optimize antigen binding by varying the incubation conditions (pH, ionic strength, time). For example, anionic carbohydrates may require poly-L-lysine coating of the plate (100 μg/mL overnight) to facilitate antigen binding.

References

1. Johnstone, A., and Thorpe, R. (1982) *Immunochemistry in Practice.* Blackwell, Oxford.
2. Hawkes, R., Niday, E., and Gordon, J. (1982) A dot immunobinding assay for monoclonal and other antibodies. *Anal. Biochem.* **119,** 142–147.

Chapter 31

Subclass Analysis and Purification of Monoclonal Antibodies

J. N. Wood

Department of Experimental Immunobiology, Wellcome Research Laboratories, Beckenham, Kent

Introduction

Once hybridoma lines have been established that secrete monoclonal antibodies of the desired specificity, it is useful to characterize the antibody subclass before mass-producing the antibody (*1,2*). IgG isotypes may be separated by chromatography on protein A–sepharose, and such affinity columns are useful for concentrating as well as purifying antibody from tissue culture supernatants. When the isotype has been determined, using commercial isotype-specific antisera conjugated to different fluorochromes and monoclonal antibodies of different isotype and specificity, double indirect immunofluorescence can allow similarities in antigen distribution to be assessed. Ideally, antibody should be purified by affinity chromatography using insolubilized antigen columns (*see* Chapter 3), although such a technique obviously requires adequate supplies of purified antigen. For many practical purposes, a crude globulin fraction from ascitic fluid is quite adequate, however.

This chapter describes a rapid simple subclass determination, concentration and purification of antibodies, and the preparation of antibody fragments, for use where possible nonspecific binding may create problems of interpretation.

The method of subclass determination described is an obvious development from the nitrocellulose dot-blot technique described in the previous chapter. By "dotting" dilute samples of isotype specific antisera on nitrocellulose sheets, and then immersing them in hybridoma supernatants, the subclass of the antibody can be detected by addition of a third radiolabelled antispecies antibody. Figure 1 shows a typical analysis of six different hybridoma supernatants. By incorporating an antiglobulin antibody on the sheets, one can also be sure that the hybridoma is still producing antibody. The method is far superior to the conventional double agar diffusion test in sensitivity, economy of reagents, simplicity, and speed.

Sodium sulfate may be used to concentrate and purify the globulin fraction from ascitic fluid. If the antibody is of the G1, G2a, or G2b subclass, a more effective purification may be obtained using protein A affinity chromatography, however.

Staphylococcus aureus (Cowan strain) has a surface protein (protein A) that binds the Fc portion of IgG. The different domain structure of the IgG subclasses is reflected in different affinities for protein A, which can be insolubilized on a sepharose matrix. By sequential elution of such an affinity column at different pHs, protein A-bound IgG can thus be purified into different subclasses. The protocol described is suitable for mouse immunoglobulins.

In order to ascribe binding to the variable region of antibody, and discount potential artifacts caused by nonspecific binding, it may be useful to produce a $F(ab')_2$

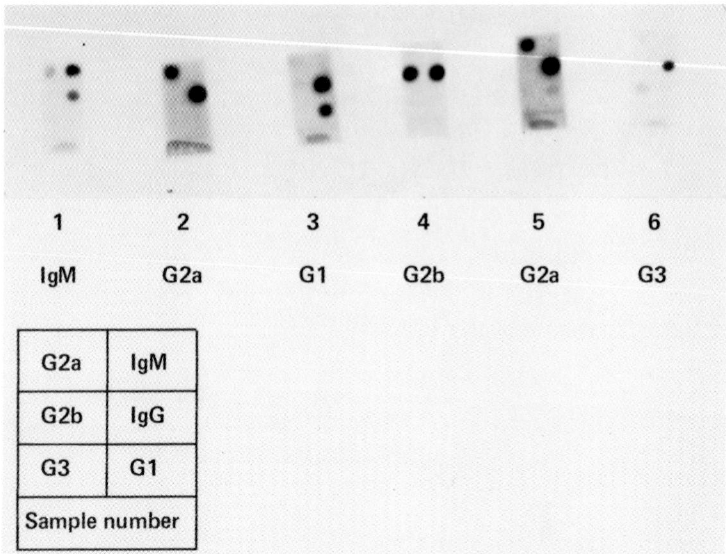

Fig. 1. Subclass determination: 1 mL samples of hybridoma supernatants were analyzed. The dark spots correspond to the antibody subclass marked on the grid. The samples shown were autoradiographed for 3 d, but a clear answer could be obtained after overnight autoradiography or the use of enzyme linked reagents (*see* Chapter 30).

fraction. This immunoglobulin fragment is prepared by pepsin hydrolysis, which cleaves the molecule distal to heavy chain disulfide bonds, so that a bivalent antibody fragment—F(ab')$_2$—is left while the constant (Fc) region of the antibody is degraded.

Materials

Subclass Determination

1. Nitrocellulose sheets.
2. Commercial antispecies anti-isotype antisera and antiglobulin antisera: e.g., rabbit anti-mouse anti G1, G2a, G2b, G3, M, and A. Store at 4°C in 0.02% azide.
3. ^{125}I-antispecies antibody (specific activity > 10 μCi/μg) preferably F(ab')$_2$ (*see* Chapter 30).

Globulin Concentration

1. 27% Na_2SO_4. Store at room temperature in water.
2. Phosphate-buffered saline (PBS) (*see* Chapter 29).

Protein-A Purification

1. Protein-A–sepharose column, 5 mL commercial column
2. 0.1*M* Phosphate buffer, pH 8.0 (store at 4°C).
3. Sodium citrate buffers, 20 mL each of:

100 m*M*	pH 6.0
100 m*M*	pH 4.5
100 m*M*	pH 3.5
100 m*M*	Citric acid

   Store frozen.
4. 2*M* Trizma base in water. Store frozen.

Preparation of $F(ab')_2$ Fragments

1. 0.1*M* sodium acetate, pH 4.5.
2. Pepsin, made up fresh from 10 mg/mL in 0.14*M* sodium acetate, pH 4.5.
3. 2*M* Trizma base in water.
4. Sephadex G-200 column (20 × 1 cm) equilibrated in PBS (*see* Chapter 28).

Purification of IgM

1. Sepharose, 4B column (50 × 2.5 cm) equilibrated in PBS.

Methods

Subclass Determination

1. Dot 1/100 dilutions of commercial anti-isotype antisera in a marked pattern on nitrocellulose sheets and air dry.

2. The sheets may be stored dry at −20°C indefinitely.
3. When samples are available, block the grids in 1 mL 5% hemoglobin in PBS in a bijou on a spiramix.
4. Quickly transfer the grids to Universals containing 1 mL tissue culture fluid and rotate for 1 h at room temperature.
5. Wash the grids three times in large volumes of PBS (5 min/wash).
6. Add 10^5 cpm $F(ab')_2$ ^{125}I-antispecies antibody (10 μCi/μg) in 1% hemoglobin in PBS for 1 h at room temperature.
7. Wash four times in PBS.
8. Dry, autoradiograph overnight at −70°C, and develop.

Globulin Concentration

1. Add two volumes of 27% sodium sulfate to 1 vol of ascitic fluid at room temperature with stirring.
2. Leave the solution for 1 h at 37°C.
3. Centrifuge at 10,000*g*, 10 min.
4. Resuspend the pellet in a volume of PBS equivalent to the original volume of ascites.
5. Dialyze exhaustively against PBS.
6. Store frozen or freeze dried.

Protein-A Affinity Chromatography

1. Pack a 1 mL protein-A Sepharose column for every 10 mg of mouse IgG in your sample.
2. Equilibrate the column with 0.1*M* phosphate buffer, pH 8.0, at 4°C.
3. Apply sample in the same buffer at 1 mL/min or less.
4. The run-through will contain IgG3, as well as IgA, most IgM, and other proteins.
5. Sequentially elute the column with 10 mL of
   a. pH 6.0 citrate buffer for IgG1
   b. pH 4.5 citrate buffer for IgG2a
   c. pH 3.5 citrate buffer for IgG2b
   d. citric acid to clear the column.

6. Neutralize both the acidified sample eluates and the column with Trizma base as quickly as possible.

The column should be kept at 4°C in azide containing (0.05%) PBS and can be re-used indefinitely. Tissue culture monoclonal antibody from fetal calf serum containing medium should be essentially pure, whereas ascitic fluid-containing antibody may contain some contaminating host IgG.

Production of F(ab′)$_2$ Fragments

1. Dissolve 10 mg IgG in 1 mL acetate buffer.
2. Add 20 μL of pepsin (10 mg/mL in acetate buffer) and incubate overnight at 37°C.
3. Neutralize the mixture with Tris to inhibit further proteolysis.
4. Separate the products of digestion by gel exclusion chromatography on a Sephadex-G200 column equilibrated with PBS.
5. The first major peak is the F(ab′)$_2$ material. A shoulder of undigested IgG may run ahead of the main peak if digestion was incomplete, and this material should be discarded.

Purification of IgM

IgM monoclonal antibodies may best be purified from concentrated globulin extracts by gel exclusion chromatography on Sepharose-4B columns. The size (~900,000 daltons) of IgM is reflected in its elution position as the first major peak.

1. Concentrate the antibody by sodium sulfate precipitation.
2. Resuspend the precipitate in a minimal volume of PBS, and apply to a Sepharose-4B column.
3. Collect the first peak of OD_{280}-absorbing material.

Notes

Although antisera are commonly stored at 4°C, in our experience a number of monoclonal antibodies stored in tissue culture medium at this temperature have proved to be unstable. It is safer to store the antibodies aliquoted at −20°C, and discard unfrozen material after a single use. A number of textbooks claim that IgM antibodies are more stable at 4°C than −20°C, but this is not true for many of the monoclonal IgMs we have studied.

References

1. Johnstone, A. P., and Thorpe, R. (1982) *Immunochemistry in Practice,* Blackwell, Oxford.
2. Ey, P. L., Prowse, S. J., and Jenkin, C. P. (1978) Isolation of pure IgG1, IgG2b immunoglobulins from mouse serum using protein-A sepharose. *Immunochemistry* **15,** 429–438.

Chapter 32

The Production of Antisera

G. S. Bailey

Department of Chemistry, University of Essex, Colchester, Essex, England

Introduction

Suitable antisera are essential for use in all immunochemical procedures. Three important properties of an antiserum are avidity, specificity, and titer. The avidity of an antiserum is a measure of the strength of the interactions of its antibodies with an antigen. The specificity of an antiserum is a measure of the ability of its antibodies to distinguish the immunogen from related antigens. The titer of an antiserum is the final (optimal) dilution at which it is employed in the procedure; it depends on the concentrations of the antibodies present and on their affinities for the antigen. The values of those parameters required for a particular antiserum very much depend on the usage to which the antiserum will be put. For example, for use in radioimmunoassay, it is best to have a monospecific antiserum of high avidity, whereas for use in immunoaffinity chromatography the monospecific antiserum should not possess too high an avidity otherwise it may prove impossible to elute the desired antigen without extensive denaturation.

A substance that, when injected into a suitable animal, gives rise to an immune response is called an immunogen. The immunogenicity of a substance is dependent on many factors, such as its size, shape, chemical composition, and structural difference from any related molecular species indigenous to the injected animal. Normally, cellular (particulate) materials are very immunogenic and induce a rapid immune response. However, the resultant antisera do not usually possess a high degree of specificity and do not store well (*1*).

Soluble immunogens differ widely in their ability to produce an immune response. In general, polypeptides and proteins of molecular weight above 5000 and certain large polysaccharides can be effective immunogens. Smaller molecules, such as peptides, oligosaccharides, and steroids, can often be rendered immunogenic by chemical coupling to a protein that by itself will produce an immune response (*2*). For most situations it is best to use the most highly purified sample available for injecting into the animal. Furthermore, it is usual to inject a mixture of the potential immunogen and an adjuvant that will stimulate antibody production.

Classically, the immune response is described as occurring in two phases. Initial administration of the immunogen induces the primary phase (response) during which only small amounts of antibody molecules are produced as the antibody producing system is primed. Further administration of the immunogen results in the secondary phase (response) during which large amounts of antibody molecules are produced by the large number of specifically programmed lymphocytes. In practice, the time-scale and nature of the immunization process does not lead to a clear recognition of two distinct phases.

All of the factors that influence antibody production have not yet been elucidated. Thus the raising of an antiserum is, to some extent, a hit or miss affair. Individual animals can respond quite differently to the same process of immunization and thus it is best to use a number of animals. The species of animal chosen for immunization will depend on particular circumstances, but, in general, rabbits are often used.

Many different methods of producing antisera have been described, varying in amount of immunogen required (often milligram quantities), route of injection, and frequency of injection (*3*). This chapter will describe a method of antiserum production (*4*) that has been successfully utilized in the author's laboratory using small doses (microgram quantities) of soluble protein as immunogen (*5*).

Materials

1. Six rabbits each of 2 kg body weight. Various types can be used, e.g., New Zealand whites, Dutch, and so on.
2. Solution of the purified immunogen in an appropriate buffer to maintain its stability.
3. Complete and incomplete Freund's adjuvant, available from various commercial sources.
4. Heat lamp.

Method

The following procedure is carred out for each rabbit in turn.

1. Thoroughly mix one volume of the immunogen solution with three volumes of complete Freund's adjuvant with the aid of a glass pestle and mortar. Initially the mixture is very viscous, but after about 5 min the viscosity becomes less. The mixture can then be transferred to a syringe. The syringe is emptied and refilled with the mixture a number of times resulting in the formation of a stable emulsion that can be injected intradermally into the prepared rabbit. The emulsion should be used within 1 h of preparation.
2. The rabbit is prepared by cutting away the long hair along the center of its back. The short hair is removed by shaving. The emulsion of immunogen and complete Freund's adjuvant is injected via a 1 mL syringe

plus 21 gage needle into two rows of five sites equidistantly spaced along the rabbit's back, each row being about 2 cm from the backbone, such that each site receives 0.1 mL of emulsion containing 1–10 μg immunogen, i.e., a total dose of 10–100 μg immunogen/rabbit.

3. After 8–10 wk, a test bleeding is carried out on the rabbit. The fur on the back of one ear is removed by shaving. The eyes of the rabbit are protected while the shaven ear is heated for less than 1 min with a heat lamp to expand the vein. The expanded vein just below the surface of the back of the ear is nicked with a scalpel blade. Blood is collected in a glass vessel (up to 20 mL can be collected in 10 min) removing the clot from the puncture wound by rubbing the ear with cotton wool from time to time. The blood is allowed to clot standing at room temperature for a few hours. The serum is separated from the clot by centrifugation and can be stored at 4°C in the presence of 0.1% sodium azide as antibacterial agent until tested.
4. If on testing the serum shows the characteristics required for its particular usage, then further bleeding can be carried out. Up to three bleedings can be made on successive days. After that it is best to allow the rabbit to rest for about a month before further bleeding.
5. If the original antiserum is unsatisfactory or if the quality of the bleedings taken over a period of several weeks or months starts to become unsatisfactory, then the rabbit can be boostered, i.e., receive a second injection of immunogen.
6. For the booster injection, half the original dose of immunogen is administered in incomplete Freund's adjuvant. The emulsion is prepared as detailed before, but it is injected *subcutaneously* into the rabbit (for example, into the fold of skin of the neck).
7. After 10 d a test bleeding is obtained from the rabbit and the serum so produced is analyzed.
8. Further bleedings can be carried out over a period of time if the boostered antiserum is satisfactory.
9. If after boostering, the antiserum is not of the desired quality it is best to disregard that rabbit. Hopefully,

one or more of the other rabbits in the group will have produced good antisera either directly or after boostering. However, in some cases, particularly with weak immunogens, it may be necessary to repeat the process of immunization with a new group of rabbits or other animals.

10. Each sample of antiserum can now be tested for its ability to form an immune precipitate with the immunogen by carrying out immunodiffusion and immunoelectrophoresis. For example, the specificity of the antibodies can be determined by running the antiserum against the immunogen and related antigens in Ouchterlony double diffusion (*see* Chapter 33). The titer and a measure of the avidity of the antiserum can be obtained by radioimmunoassay (*see* Chapter 37).

Notes

1. Many antisera can be satisfactorily stored at 4°C in the presence of an antibacterial agent for many months. After some time, the antiserum solution may become turbid and even contain a precipitate (mostly of denatured lipoprotein). Even so, there should be no significant reduction in the quality of the antiserum. If necessary, the solution can be clarified by membrane filtration. For prolonged storage, the antiserum can be kept at −20°C in small quantities so as to avoid repeated thawing and freezing.
2. Animals other than rabbits can be used for immunization, e.g., mice, rats, guinea pigs, sheep, goat, and so on. The rabbit is often a good initial choice, but if results are unsatisfactory, other species can be tried. Obviously only small volumes of antisera can be generated in the small species, whereas large volumes can be obtained from the larger species. The latter though do require more immunogen and are more expensive to maintain.
3. A clean sample of antiserum should be straw-colored. Pink coloration is due to partial hemolysis, but should not affect the properties of the antiserum.

4. The dose of immunogen employed can be of crucial importance in many procedures for antibody production. A state of tolerance can be induced in the animal with little or no production of antibody if too much or too little immunogen is repeatedly given over a relatively short period of time. The method described in this chapter should not suffer from that effect since there is a gap of at least 10 wk between the initial and booster injections. In general, the lower the dose of antigen the greater is the avidity of the antiserum.
5. Since the antisera produced by convential methods consist of mixtures of different antibody molecules, it is to be expected that the properties of the antisera collected during the prolonged period of immunization may change. Thus each bleeding should be tested for specificity, titer, and avidity.

References

1. Hurn, B. A. L., and Chantler, S. M. (1980) Production of Reagent Antibodies in *Methods in Enzymology* **70** (eds. Van Vunakis, H., and Langone, J. J.) pp. 104–141. Academic Press, New York.
2. Erlanger, B. F. (1980) The Preparation of Antigenic Hapten-Carrier Conjugates: A survey, in *Methods in Enzymology* **70** (eds. Van Vunakis, H., and Langone, J. J.), pp. 85–104. Academic Press, New York.
3. Weir, D. M. (1978) *Handbook of Experimental Immunology* **3**, third edition. Blackwell, Oxford.
4. Vaitukaitis, J. L. (1981) Production of Antisera with Small Doses of Immunogen: Multiple Intradermal Injections in *Methods in Enzymology* **73** (eds. Langone, J. J., and Van Vunakis, H.) pp. 46–57. Academic Press, New York.
5. Hussain, M., and Bailey, G. S. (1982) An improved method of isolation of rat pancreatic prokallikrein. *Biochim. Biophys. Acta* **719**, 40–46.

Chapter 33

Immunodiffusion in Gels

G. S. Bailey

Department of Chemistry, University of Essex, Colchester, Essex, England

Introduction

Immunodiffusion in gels encompasses a variety of techniques that are useful for the analysis of antigens and antibodies (*1*). The fundamental immunochemical principles behind their use are exactly the same as those that apply to antigen–antibody interactions in the liquid state. Thus an antigen will rapidly react its specific antibody to form a complex, the composition of which will depend on the nature, concentrations, and proportions of the initial reactants. As increasing amounts of a multivalent antigen are allowed to react with a fixed amount of antibody, precipitation occurs, in part because of extensive crosslinking between the reactant molecules. Initially the antibody is in excess and all of the added antigen is present in the form of an insoluble antigen–antibody aggregate. Addition of more antigen leads to the formation of more immune precipitate. However, a point is reached beyond which further addition of antigen produces an excess of antigen and leads to a reduction in the amount of the precipitate (*see*

Fig. 1) because of the formation of soluble antigen–antibody complexes. The analysis of such interactions occurring in gels is of much higher sensitivity and resolution than that for the liquid state, thus explaining the extensive use of immunochemical gel techniques.

The methods employing immunodiffusion in gels are often classified as simple (single) diffusion or double diffusion. In simple diffusion, one of the reactants (often the antigen) is allowed to diffuse from solution into a gel containing the corresponding reactant, whereas in double diffusion both antigen and antibody diffuse into the gel. A variety of information, both qualitative and quantitative, can be obtained from the numerous techniques (2). This chapter will be concerned with two methods that are very widely employed for immunochemical analysis.

Ouchterlony Double Immunodiffusion

This technique is often used in qualitative analysis of antigens and antisera (2).

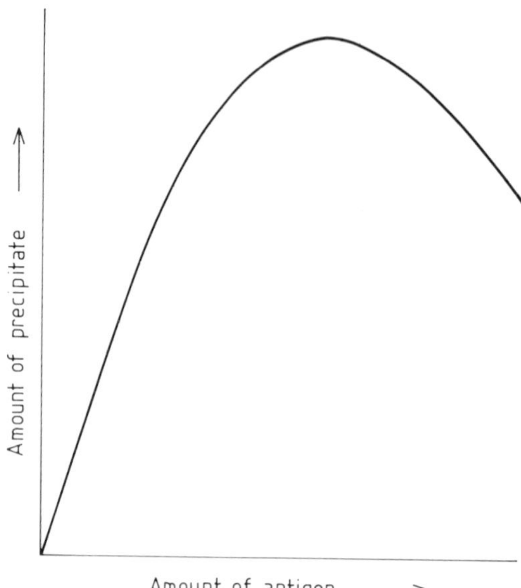

Fig. 1. Variation in the amount of immune precipitate on addition of increasing amounts of antigen to a fixed amount of antibody.

Single Radial Immunodiffusion (SRID)

This method is used for the quantitative analysis of antigens (3).

Materials

1. Agarose.
2. 0.07*M* barbitone buffer, pH 8.6, containing 0.01% thimerosal as antibacterial agent. The buffer is prepared by dissolving sodium barbitone (14.5 g), disodium hydrogen phosphate decahydrate (7.16 g), boric acid (6.2 g), and sodium ethyl mercurithiosalicylate (0.1 g) in distilled water and making the final volume up to 1 L.
3. Solutions of antigen and antiserum.
4. Plastic or glass Petri dishes or rectangular plates.
5. Gel punch and template. Suitable gel punches of various sizes and templates of various designs can be obtained from commercial sources.
6. Flat level surface.
7. Humidity chamber at constant temperature.
8. 0.1*M* sodium chloride in distilled water.
9. Staining solution. The solution is prepared by mixing ethanol (90 mL), glacial acetic acid (20 mL), and distilled water (90 mL), and then adding Coomassie Brilliant Blue R-250 (1 g). The solution can be re-used several times.
10. Destaining solution. The composition is the same as that of the staining solution but without the dye.

Method

Ouchterlony Double Immunodiffusion

1. Agarose (1 g) is dissolved in the barbitone buffer (100 mL) by heating to 90°C on a water bath with constant stirring.

2. The agarose solution is poured to a depth of 1–2 mm into the Petri dishes or onto the rectangular plates that had previously been set on a horizontal level surface. The gels are allowed to form on cooling and when set (5–10 min) can be stored at 4°C in a moist atmosphere for at least 1 wk.
3. A template of the desired pattern, according to the number of samples to be analyzed, is positioned on top of the gel. Commonly used patterns are shown in Fig. 2. The gel punch, which is connected to a water vacuum pump, is inserted into the gel in turn through each hole of the template so that the wells are cleanly formed as the resultant agarose plugs are sucked out.
4. The wells are filled with the solutions of antigen and antisera until the meniscus just disappears. The concentrations of antigen solutions and the dilution of the antiserum to be used have to be established largely by trial and error by running pilot experiments with solutions of different dilutions. However, as a rough guide, for the analysis of antigens, the following concentrations of antigen solutions can be run against undiluted antiserum: for a pure antigen, 1 mg/mL; for a partially pure antigen, 50 mg/mL, for a very impure antigen, 500 mg/mL, using 5 μL samples of antigen solutions and antiserum.
5. The gel plate is then left in a moist atmosphere, e.g., in a humidity chamber *at a constant temperature* of 20°C for 24 h.

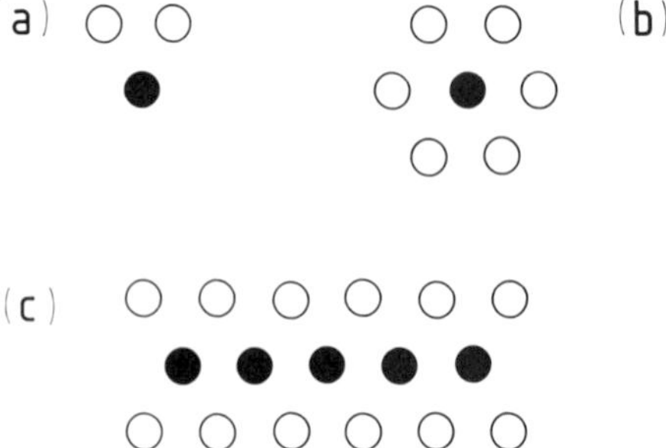

Fig. 2. Patterns of wells often used in double immunodiffusion: ○, well contains antigen; ●, well contains antiserum.

6. The precipitin lines can then be recorded *either* directly by photography with dark-field illumination or by drawing the naked eye observation on suitable oblique illumination of the plate against a dark background *or* after staining. Prior to staining, excess moisture is removed from the gel plate by application of a 1 kg weight (e.g., a liter beaker full of water) to a wad of filter paper placed on top of the gel for 15 min. Soluble protein is removed by washing the gel (3 × 15 min) in 0.1*M* sodium chloride solution followed by further pressing. The gel is dried using cold air from a hair dryer and is then placed in the staining solution for 5 min. The plate is finally washed with distilled water and placed in the destaining solution for about 10 min.

Single Radial Immunodiffusion (SRID)

1. Agarose (2 g) is dissolved in the barbitone buffer (100 mL) by heating to 90°C on a water bath with constant stirring.
2. The solution is allowed to cool to 55°C and is mixed with an equal volume of a suitable dilution of the monospecific antiserum also at 55°C. The optimal dilution of the antiserum has to be found from pilot experiments using different dilutions of the antiserum. The dilution chosen depends in part on the range of antigen concentrations that is required to be measured.
3. The agarose solution containing the antiserum is poured onto the rectangular plates that had previously been set on a horizontal level surface. The gels are allowed to form on cooling and the gel plates are normally used within 24 h, but can be stored for longer periods at 4°C provided dessication is avoided.
4. Using a gel punch and a template, a row of the desired number of wells of diameter 2 mm is cut into the gel.
5. At least three of the wells are filled with standard solutions of antigen of known concentration. The other wells are filled with the solutions containing the antigen at unknown concentrations. For the most accurate work it is necessary to add all samples via a microsyringe.

6. The gel plate is then left in a humidity chamber for an appropriate period of time. As the antigen diffuses into the gel containing the antiserum a disc of immune precipitate is formed. The final, maximum area of that disc is directly proportional to the initial concentration of the antigen in the well (and inversely proportional to the antibody concentration in the gel). The time required to achieve maximum area of precipitation depends on the velocity of diffusion of the antigen. That, in turn, is dependent on temperature and molecular size of the antigen. In general, the diffusion at room temperature must be allowed to continue for a few days, taking measurements of the areas of the discs every 24 h until no further increase takes place.
7. The area of each disc is measured in terms of its diameter, which can be measured directly with the aid of a magnifying glass on suitable oblique illumination of the plate positioned over a dark background. For increased resolution of the discs, the plate can be placed in 1% tannic acid for 3 min prior to viewing.
8. At the end of the diffusion process the plate can be stained. Soluble protein is removed from the gel by washing for 2 d with several changes of 0.1*M* sodium chloride solution. The plate is then pressed, dried, stained and destained as detailed for Ouchterlony double immunodiffusion (*see* above).
9. A standard graph is constructed by plotting the diameters of the discs against the logarithm of the antigen solutions of known concentration. The concentrations of the antigen in the test samples can then be determined by simple interpolation (*see* Fig. 3).

Notes

1. Ouchterlony double diffusion is frequently used for comparing different antigen preparations.

   If the antigen solution contains several different antigens that can react with the antibodies of the antiserum, then multiple lines of precipitation will be produced. The relative position of each line is determined by the local concentration of each antigen and antibody

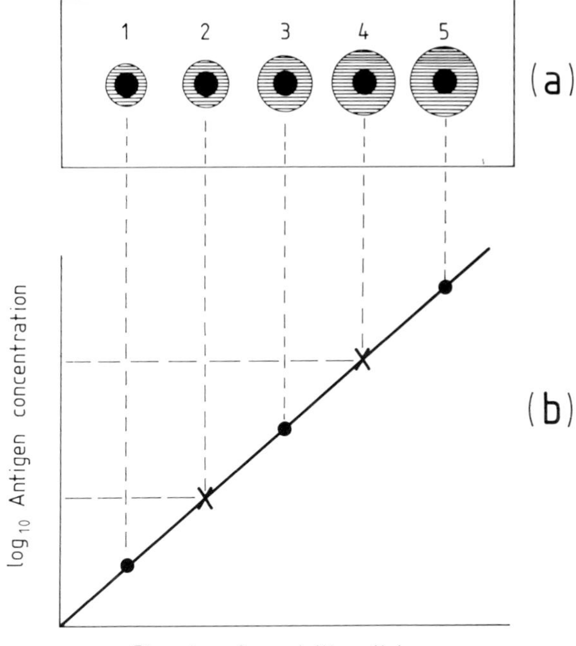

Fig. 3. Single radial immunodiffusion. (a) Precipitin discs formed at termination of diffusion of antigen into gel containing monospecific antiserum. Wells 1, 3, and 5 contained standard antigen solutions of increasing concentration. Wells 2 and 4 contained samples of the antigen at unknown concentrations. Shaded area, immunoprecipitate. (b) Semilog plot of diameter of the precipitin discs of standard antigen solutions (●) against concentration. Measurement of the diameters of the precipitin discs of the unknown solutions (×) allows an estimation of the antigen concentration to be made by simple interpolation.

in the gel. In turn, those local concentrations depend not only on the initial concentrations of the reactants in the wells, but on the rates at which they diffuse through the gel and hence are also dependent on molecular size.

If different antigen preparations, each containing a single antigenic species capable of reacting with the antiserum used, are allowed to diffuse from separate wells, then the degree of similarity of the antigens can be assessed by observation of the geometrical pattern produced. For example, Figure 4 shows the four basic precipitin patterns that can be produced in a balanced system by two related antigens interacting with an antiserum that contains antibodies that can recognise both antigens.

Pattern (a) is called the "pattern of identity" or "pattern of coalescence" and indicates that the antibodies in the antiserum employed cannot distinguish the two antigens, i.e., the two antigens are immunologically identical as far as that antiserum is concerned.

Pattern (b) is called the "pattern of non-identity" or "pattern of absence of coalescence" and indicates that none of the antibodies in the antiserum employed react with antigenic determinants that may be present in both antigens, i.e., the two antigens are unrelated as far as that antiserum is concerned.

Pattern (c) is called a "pattern of partial identity" or "pattern of partial coalescence of one antigen" and indicates that more of the antibodies in the antiserum employed react with one antigen (that diffusing from the left-hand well in Fig. 4c) than the other antigen. The "spur", extending beyond the point of partial coalescence, is thought to result from the determinants present in one antigen but lacking in the other antigen.

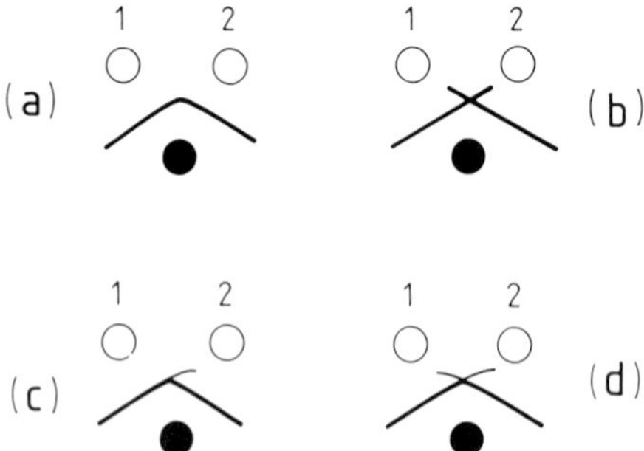

Fig. 4. Patterns of precipitin lines: (a) pattern of coalescence; (b) pattern of absence of coalescence; (c) pattern of partial coalescence of one antigen; (d) pattern of partial coalescence of two antigens. Well 1 contains antigen 1; well 2 contains antigen 2; well ● contains antiserum;—precipitin line.

Pattern (d) is called "the pattern of partial coalescence of two antigens" and is another type of "pattern of partial identity." It indicates that some of the antibodies in the antiserum employed react with both antigens, whereas other antibodies only react with one or other of the antigens. Thus the two antigens have at least one antigenic determinant in common, but also differ in other antigenic determinants.

2. The following comments relate to Ouchterlony Double Immunodiffusion.

   (a) If glass petri dishes or rectangular plates are used for this technique, it normally is necessary for them to be precoated with a 0.5% agarose gel prior to formation of the main gel to prevent the main gel from becoming detached during the washing and staining procedures.
   (b) Buffers of different composition and pH from the described buffer can be used if more convenient, e.g., 0.05*M* phosphate buffer, pH 7.2, containing 0.1*M* sodium chloride.
   (c) The agarose gel solution, once made, can be separated into test tubes and stored at 4°C for several weeks. Approximately 12 mL of 1% agarose gel solution is required for an area of 75 mL. The contents of each test tube can be simply melted as required.
   (d) Diffusion of antigens and antibodies is more rapid at higher temperatures. Thus double immunodiffusion is often run for 48 h at 4°C, or 24 h at room temperature or for as little as 3 h at 37°C. It is important to maintain a constant temperature otherwise artifacts may be produced.
   (e) Other artifacts can be produced by refilling of the wells or by denaturation of the antigen during diffusion. If the latter is a possibility, it is best to run the immunodiffusion at 4°C.
   (f) The sensitivity of the Ouchterlony technique is largely determined by the relative concentrations of the antigens and antibodies, and by the separation of the wells. In the latter case, the closer the wells the greater is the sensitivity. It should be

possible to detect protein solutions of 10 μg/mL concentration.

3. The following comments relate to Single Radial Immunodiffusion.
   (a) If samples in neighboring wells are too concentrated, their precipitin discs will overlap and in that case the samples have to be run again at higher dilutions.
   (b) If the antiserum employed is not monospecific for the required antigen in an impure sample a number of precipitin discs will be produced by the different antigens interacting with their corresponding antibodies. The problem then is to identify the disc produced by the antigen under study. One solution is to rerun each sample in duplicate with one well containing just the sample and the duplicate well containing the sample plus a fixed amount of standard antigen. The disc caused by the antigen under study can then be identified by its larger area around the duplicate well.
   (c) For maximum sensitivity the antiserum used in the gel should be diluted with non-immune serum, i.e., serum from a non-immunized animal. Furthermore, the dimensions of the wells and volumes of samples and standards applied can be increased. It should be possible to measure protein solutions of 1 μg/mL concentration.

References

1. Oudin, J. (1980) Immunochemical analysis by antigen-antibody precipitation in gels in *Methods in Enzymology* **70** (eds. Van Vunakis, H., and Langone, J. J.) pp. 166–198. Academic Press, New York.
2. Ouchterlony, O. (1968) *Handbook of Immunodiffusion and Immunoelectrophoresis*. Ann Arbor Science Publications, Michigan.
3. Vaerman, J. P. (1981) Single radial immunodiffusion, in *Methods in Enzymology* **73** (eds. Langone, J. J., and Van Vunakis, H.) pp. 291–305. Academic Press, New York.

Chapter 34

Crossed Immunoelectrophoresis

Graham B. Divall

Metropolitan Police Forensic Science Laboratory, London, England

Introduction

Crossed immunoelectrophoresis is a simple, quick, and sensitive technique for the qualitative detection of a wide range of protein antigens.

The method consists of making two small wells, approximately 5 mm apart, in a thin layer of agar buffered at pH 8.6. One of the wells (sample well) is filled with a solution thought to contain the protein in question. The second (antiserum) well is filled with a monospecific antiserum against that protein. The whole gel is then subjected to electrophoresis for about 15 min. At pH 8.6, the vast majority of proteins are negatively charged and migrate towards the anode. The antibodies, contained in the γ-

globulin fraction of the antiserum, have a relatively low negative charge, but their net movement is towards the cathode as a result of electroendosmosis. Protein antigen and corresponding antibody are therefore caused to migrate towards each other in the gel and form lines of precipitation between the wells.

The phenomenon of electroendosmosis is caused by the presence of negatively charged groups in the agar gel and the surface adsorption of positive ions from the buffer. The agar gel as the supporting medium is, of course, fixed, but the adsorbed positive ions migrate towards the cathode. This accelerates the migration of cations towards the cathode and retards or even reverses the migration of anions towards the anode. In the case of crossed immunoelectrophoresis, the net effect of electrophoresis and electroendosmosis is for the γ-globulins to migrate towards the cathode.

The method is similar to simple gel immunodiffusion. However, it is quicker and more sensitive because all the antigen and antibody are driven towards each other under the influence of the electric field rather than being allowed to diffuse radially.

The technique is suitable for the detection of any protein with an electrophoretic mobility at pH 8.6 greater than γ-globulin and for which a monospecific precipitating antiserum is available. It is widely used to determine the species origin of bloodstains in forensic science and in many clinical screening procedures, for example, the detection of Australia Antigen.

Materials

1. Agar gel buffer, pH 8.6, consisting of: 7.0 g sodium barbiturate; 1.1 g diethyl barbituric acid; 1.02 g calcium lactate dissolved in 1 L of distilled water.
2. Electrophoresis tank buffer, pH 8.6, consisting of: 17.52 g sodium barbiturate; 2.76 g diethyl barbituric acid; and 0.77 g calcium lactate dissolved in 2 L of distilled water.
3. Staining solution, consisting of: 0.4 g naphthalene black 10 B; 200 mL methyl alcohol; 40 mL glacial acetic acid; and 200 mL distilled water.

4. Wash/destain solution, consisting of 500 mL methyl alcohol; 100 mL glacial acetic acid; and 500 mL distilled water.
5. 2% w/v stock agar in distilled water: Prepared by mixing 10 g dry agar powder with 500 mL distilled water in a 2 L glass beaker. Heat the suspension over a low Bunsen flame and stir continuously with a glass rod. Bring to the boil and continue boiling for a few minutes to ensure that the agar is completely dissolved. Tip the hot agar solution into a storage container (plastic sandwich boxes are ideal) and allow to cool and gel. Fit a lid and store 2% stock agar gel at 4°C.
6. Glass plates (microscope slides) 7.5 × 5 cm, approximately 1.5 mm thick.

Method

1. Take 50 g stock 2% agar and dissolve in 50 mL gel buffer by gentle heating in a boiling water bath or over a low Bunsen flame.
2. Take 12 glass plates. Scribe an identification mark in one corner of each plate with a diamond pencil. Clean the surfaces with a tissue moistened with methyl alcohol and place the slides on a horizontal surface.
3. Carefully transfer 7 mL of the hot agar solution to the surface of each glass plate. This is best achieved by using a large plastic hypodermic syringe. Leave the plates for 10 min to allow the agar to cool and gel, then place them in a moist chamber (plastic box with a tight fitting lid and lined with damp filter paper) and store at 4°C.
4. Make a series of paired wells in the gel. They should be approximately 1–1.5 mm in diameter and 5–8 mm apart. The wells are made and plugs of agar removed by means of a glass Pasteur pipet or a piece of metal tubing attached to a vacuum pump.
5. Place the gel plate on a dark surface with the identification mark in the top right-hand corner. For each pair of wells, fill the one on the left with a solution of the sample and the one on the right with antiserum. This can be difficult and is best done using finely

drawn Pasteur pipets. Be careful not to overfill the wells. For maximum reproducibility, use a microsyringe and fill the wells with 2–3 μL of sample solution or antiserum.

6. Invert the gel plate (the solutions will be retained in their wells) and place it across the bridge of an electrophoresis tank with the cathode on the left and the anode on the right. The edges of the gel surface should rest on filter paper wicks to form a bridge gap of about 6.5 cm. The identification mark should now be visible in the bottom right-hand corner. The sample well of each pair will be nearest the cathode and the antiserum well nearest to the anode.
7. Turn on the power pack and subject the gel plate to electrophoresis at 130 V (20 V/cm) for 15 min.
8. After electrophoresis, remove the plate from the tank and view the gel with oblique lighting. A positive result is indicated by a clear sharp white precipitin line between the two wells. Very weak results cannot be seen at this stage. Furthermore, for some samples such as those encountered in forensic analysis, a nonspecific protein precipitate forms between the wells and can be confused with a true immunoprecipitate. For these reasons, the gel is best washed and stained.
9. Place the gel plate in a 1*M* saline bath for at least 24 h. During this time all protein will be washed out of the gel except for immunoprecipitates that have become enmeshed in the gel matrix. Transfer the gel plate to a distilled water bath for 2 h to remove the salt. The washing procedures are more efficient if the saline/water baths are shaken gently.
10. Remove the gel plate from the water and cover the gel surface with a piece of wet filter paper. The plate can now be left overnight at room temperature, or placed in a incubator at 50–60°C for several hours to allow the gel to dry.
11. When the gel is dry the filter paper is removed. Take care here because the paper tends to stick to the surface of the dried gel and if it is pulled off, the gel film can be torn and lost. To avoid this, moisten the paper with tap water and then gently peel it away from the

gel film. The dried gel will appear as a shiny film fixed to the glass plate. Wash the plate under running water while gently rubbing the gel surface with a finger to remove fragments of adherent paper.

12. Place the plate in the staining solution for 15 min. Shaking is unnecessary. Transfer the plate to the wash/destain solution for 10 min. Destaining is more efficient if the bath is gently shaken.
13. Remove the plate, allow it to dry, and then view over a light bench. The immunoprecipitates appear as dark blue lines between the sample and antiserum wells against a colorless background. A typical result is shown in Fig. 1.

Notes

1. An obvious source of error occurs when the sample and antiserum wells, or the polarity of the electrophoresis tank, are reversed. In either case the reactants will be driven apart rather than together.
2. The gels sometimes have a tendency to float off the glass plate during the washing stage. This can be mini-

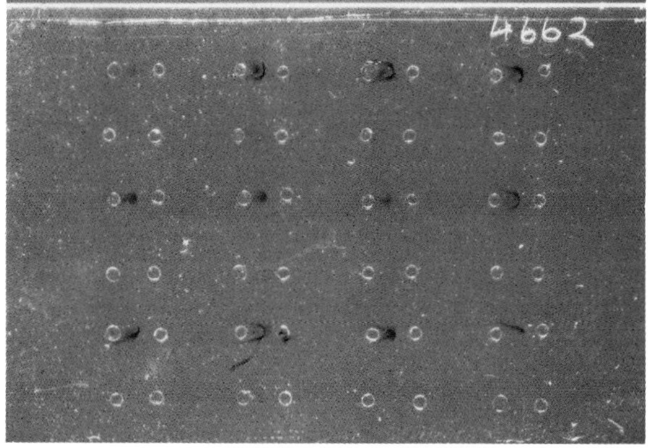

Fig. 1. A crossed immunoelectrophoresis plate used for the species identification of bloodstains in forensic science. The gel has been dried and stained. A dark (stained) precipitin line can be seen between some of the wells indicating a positive reaction between the bloodstain extract and a polyspecific antiserum to human serum proteins.

mized by precoating the glass plates with 0.2% agar. Simply brush the top surface of the plate with melted 0.2% agar, allow to dry, and proceed from step 3. Alternatively, and less bother, cut the top left hand corner of the gel off before washing. This allows the correct orientation of gel to plate to be made before drying.

3. The sensitivity of the method is ultimately dependent on the titer and avidity properties of the antiserum used. For example, different antisera have been able to detect human blood diluted to between 1 in 15,000 and 1 in 500,000.
4. The method is absolutely dependent on the agar gel exhibiting electroendosmotic flow. Different agars show different degrees of electroendosmosis and the best agar for any particular system usually has to be found by trial and error.
5. The formation of an immunoprecipitate and its position between the wells is influenced by the antigen–antibody ratio. Precipitates 'hugging' the sample or antiserum well are to be avoided and it is sometimes necessary to test dilution series of the sample against the antiserum, the sample against a dilution series of the antiserum, or both.

References

1. Culliford, B. J. (1964) Precipitin reactions in forensic problems. *Nature* **201,** 1092–1094.

Chapter 35

Rocket Immunoelectrophoresis

John M. Walker

School of Biological and Environmental Sciences, The Hatfield Polytechnic Hatfield, Hertfordshire, England

Introduction

Rocket immunoelectrophoresis (also referred to as electroimmunoassay) is a simple, quick, and reproducible method for determining the concentration of a specific protein in a protein mixture. The method, originally introduced by Laurell (*1*) involves a comparison of the sample of unknown concentration with a series of dilutions of a known concentration of the protein, and requires a monospecific antiserum against the protein under investigation. The samples to be compared are loaded side-by-side in small circular wells along the edge of an agarose gel that contains the monospecific antibody. These samples (antigen) are then electrophoresed into the agarose gel where interaction between antigen and antibody takes place. In the presence of excess antigen, the antigen–antibody complex is soluble, but as the antigen moves further into the gel, more antigen combines with antibody until a point of equivalence is reached. At this stage the antigen–antibody complex is insoluble. The end result is a precipitation 'rocket' spreading out from the loading well. Since the height of the peak depends on the relative excess of antigen over antibody, a comparison of the peak heights of the unknown and standard samples allows the unknown protein concentration to be determined. Although particularly useful for measuring the concentration of any given protein in a serum sample, the method may of course be used to measure or compare protein concentrations in any mixture (e.g., urine, cerebrospinal fluid, tissue homogenates, and so on) as long as an antiserum to the protein is available.

Materials

1. 0.06*M* barbitone buffer, pH 8.4, is prepared from sodium barbitone and barbitone. Dilute this buffer 1:1 with distilled water for use in the electrophoresis tanks.
2. Agarose, 2% w/v (aqueous).
3. Protein stain: 0.1% Coomassie Brilliant Blue in 50% methanol/10% acetic acid. (Dissolve the stain in the methanol component first, and then add the appropriate volumes of acetic acid and water). Destain: 10% methanol/7% acetic acid.
4. Glass plates, 5 × 5 cm, 1.0–1.5 mm thick.

Method

1. Melt the 2% agarose by heating to 100°C and then place the bottle in a 52°C water bath. Stand a 5 mL pipet in this solution and allow time for the agarose to cool to 52°C. At the same time the pipet will be warming.
2. Place a test-tube in the water bath and add barbitone buffer (0.06*M*, 2.8 mL). Leave this for about 3 min to

allow the buffer to warm up and then add agarose solution (2.8 mL). This transfer should be made as quickly as possible to avoid the agarose setting in the pipet. Using a pipet with the end cut off to give a larger orifice allows a more rapid transfer of the viscous agarose solution. Some setting of agarose in the pipet is inevitable, but is of no consequence. Briefly mix the contents of the tube and return to the water bath.

3. Thoroughly clean a 5 × 5 cm glass plate with methylated spirit. When dry, put the glass plate on a leveling plate or level surface.
4. Add antiserum (50 μL, but *see* Note 3) to the diluted agarose solution and *briefly* mix to ensure even dispersion of the antiserum. Briefly return to the water bath to allow any bubbles to settle out.
5. Immediately pour the contents of the tube onto the glass plate. Keep the neck of the tube close to the center of the plate and pour slowly. Surface tension will prevent the liquid running off the edge of the plate. Alternatively, tape can be used to form an edging to the plate. The final gel will be approximately 2 mm thick.
6. Allow the gel to set for 5 min and then make holes (1 mm diameter) at 0.5 cm spacings, 1 cm from one edge of the plate. This is most easily done by placing the gel plate over a predrawn (dark ink) template when well positions can easily be seen through the gel. The wells can be made using a Pasteur pipet or a piece of metal tubing attached to a weak vacuum source (e.g., a water pump).
7. Place the gel plate on the cooling plate of an electrophoresis tank and place the electrode wicks (six layers of Whatman No. 1 paper, prewetted in electrode buffer) over the edges of the gel. Take care not to overlap the sample wells and ensure that these wells are nearest the cathode.
8. Each well is now filled with 2 μL of sample. Samples should be diluted accordingly with buffer so that 2 μL Samples can be loaded. It is important to completely fill the wells, and also to ensure that all samples are

loaded in the same volume. The loading of wells should be carried out as quickly as possible to minimize diffusion from the wells. Some workers prefer to load wells with a current (1–2 mA) passing through the gel, which should overcome any diffusion problems.

9. Once the samples are loaded, electrophoresis is commenced. For fast runs, a current of 20 mA (~200 V, 10 V/cm across the plate) is passed through the gel for about 3.0 h. Alternatively, gels can be run at 2–3 mA overnight. It is important that water cooling be used to dissipate heat generated during electrophoresis.
10. At the end of the run, precipitation rockets can be seen in the gel. These are not always easy to visualize and are best observed using oblique illumination of the gel over a dark background.
11. A more clear result can be obtained by staining these precipitation peaks with a protein stain. Firstly, the antiserum in the gel, which would otherwise stain strongly for protein, must be removed. One way of doing this is to wash the gel with numerous changes of saline (0.14*M*) over a period of 2–3 d. The washed gel may then be stained directly or dried and then stained to give a permanent record. To dry the washed gel, place it on a clean sheet of glass, wrap a piece of wet filter paper around the gel, and place it in a stream of warm air for about 1 h. Wet the paper again with distilled water and remove it to reveal the dried gel fixed to the glass plate. Any drops of water on the gel can be finally removed in a stream of warm air.
12. A more convenient and quicker method of preparation is to blot the gel dry. With the gel on a sheet of clean glass, place eight sheets of filter paper (Whatman No. 1) on top of the gel and then apply a heavy weight (lead brick) for 1–2 h. After this time, carefully remove the now wet filter papers to reveal a flattened and nearly dry gel. Complete the drying process by heating the gel in a stream of hot air for about 1 min. When completely dry, the gel will appear as a glassy film.

13. Whichever method of gel preparation is used, the gel can now be stained by placing it in stain (with shaking) for 10 min, followed by washing in destain.
14. Peak heights can now be measured and a graph of concentration against peak height plotted. A typical electrophoretic run is shown in Fig. 1.

Notes

1. The size of gel described here (5 × 5 cm) is suitable for routine use. However, for highly accurate determina-

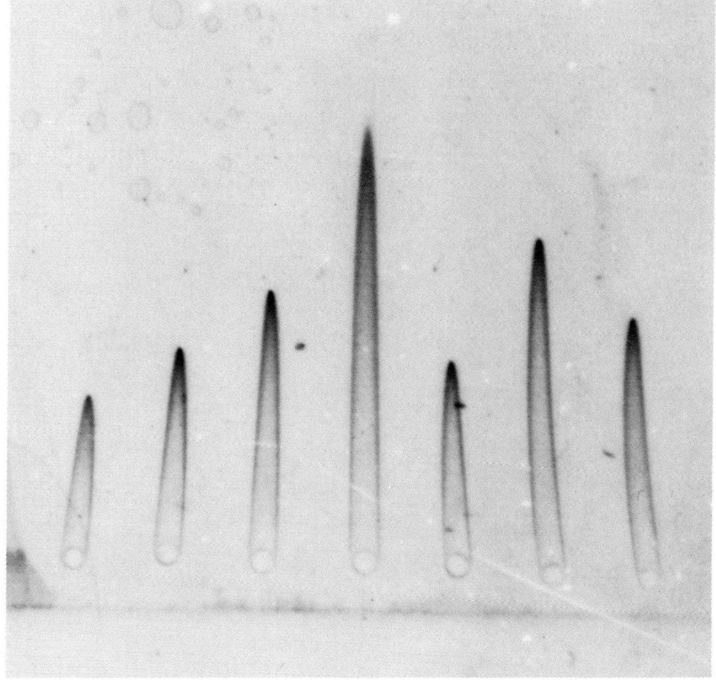

Fig. 1. A rocket immunoelectrophoresis gel (5 × 5 cm) run at 20 mA for 3h, and stained for protein with Coomassie Brilliant Blue. The gel contained anti-bovine serum albumin (50 μL, anti-BSA) and the sample loading (2 μL/well) were (left to right): 1. 40 ng BSA; 2. 67 ng BSA; 3. 100 ng BSA; 4. 200 ng BSA; 5. a 1:1200 dilution of bovine serum; 6. 130 ng BSA; 7. 100 ng BSA.

tions larger plates should be used (7.5 × 7.5 cm) with larger wells (5–10 μL volumes) so that sample loadings can be made with greater accuracy, and the correspondingly larger peaks measured more accurately. The volume of agarose and antisera used should of course be increased accordingly.

2. When possible, the gel plates used should be as thin as possible to maximize the effect of the cooling plate. This is particularly important in the case of the shorter run time when considerable heat is evolved, which can cause distortion of rockets.
3. The amount of antiserum to be used in the gel depends of course on the antibody titer, and is best determined by trial and error. The amount quoted in the Method section (50 μL) is the volume of commercially available anti-bovine serum albumin used in our laboratory when measuring bovine serum albumin levels (see Fig. 1). This is an ideal test system for someone setting up the technique for the first time.
4. Because of the thickness of the electrophoresis wicks used, they are quite heavy when wet and have a tendency to 'slip off' the gel during the course of the run. This can be prevented by placing thick glass blocks on top of the wicks where they join the gel, or by placing a heavy glass sheet across both wicks. This, by also covering the gel, has the added advantage of reducing evaporation from the gel.
5. A uniform field strength over the entire gel is critical in rocket immunoelectrophoresis. For this reason the wick should be exactly the width of the gel.
6. When setting up the electrophoresis wicks, it is important that they be kept well away from the electrodes. Since the buffer used is of low ionic strength, if electrode products are allowed to diffuse into the gel, they can ruin the gel run. Most commercially available apparatus is therefore designed such that the platinum electrodes are masked from the wicks and also separated by a baffle. To minimize these effects, the use of large tank buffer volumes (~800 mL per reservoir) is also encouraged.
7. The time quoted for electrophoresis is suitable for most samples. However, for proteins with low electro-

phoretic mobility under the conditions used, these times may not be long enough to allow complete development of rockets. The minimum time necessary for electrophoresis should be determined in initial trial experiments. Do note that it is not possible to 'overrun' the electrophoresis. Once the rockets are formed they are quite stable to further electrophoresis.

8. Rocket immunoelectrophoresis is accurate for samples with protein concentrations as low as 10 μg/mL and require as little as 20 ng of protein to be loaded onto a gel.

References

1. Laurell, C-B. (1966) Quantitative estimation of proteins by electrophoresis in agarose gel containing antibodies. *Anal. Biochem.* **15,** 45–52.

Chapter 36

Radioiodination of Proteins

G. S. Bailey

Department of Chemistry, University of Essex, Colchester, Essex, England

Introduction

Many different substances can be labeled by radioiodination. Such labeled molecules are of major importance in a variety of investigations, e.g., studies of intermediary metabolism, determinations of agonist and antagonist binding to receptors, quantitative measurements of physiologically active molecules in tissues and biological fluids, and so on. In most of those studies, it is necessary to measure very low concentrations of the particular substance and that in turn implies that it is essential to produce a radioactively labeled tracer molecule of high specific radioactivity. Such tracers, particularly in the case of polypeptides and proteins, can often be conveniently produced by radioiodination.

Two γ-emitting radioisotopes of iodine are widely available, ^{125}I and ^{131}I. As γ-emitters they can be counted directly in a well-type crystal scintillation counter (commonly referred to as a γ counter) without the need for sample preparation in direct contrast to β-emitting radionuclides, such as ^{3}H and ^{14}C. Furthermore, the count rate produced by 1 g atom of ^{125}I is approximately 75 times and 35,000 times greater than that produced by 1 g-at of ^{3}H and ^{14}C, respectively. In theory, the use of ^{131}I would result in a further sevenfold increase in specific radioactivity. However, the isotopic abundance of commercially available ^{131}I rarely exceeds 20%, because of ^{127}I and ^{129}I contaminants, and its half-life is only 8 d. In contrast, the isotopic abundance of ^{125}I on receipt in the laboratory is normally at least 90% and its half-life is 60 d. Also, the counting efficiency of a typical well-type crystal scintillation counter for ^{125}I is approximately twice that for ^{131}I. Thus, in most circumstances, ^{125}I is the radionuclide of choice for radioiodination.

Several different methods of radioiodination of proteins have been developed (*1*). They differ, among other respects, in the nature of the oxidizing agent for converting $^{125}I^{-}$ into the reactive species $^{125}I_2$ or $^{125}I^{+}$. In the main, those reactive species substitute into tyrosine residues of the protein, but substitution into other residues, such as histidine, cysteine, and tryptophan, can occur in certain circumstances.

In this chapter two methods of radioiodination of proteins will be described.

The Chloramine-T Method

This method, developed by Hunter and Greenwood (2), is probably the mostly widely used of all techniques of protein radioiodination. It is a very simple method in which the radioactive iodide is oxidized by chloramine-T in aqueous solution. The oxidation is stopped after a brief period of time by addition of excess reductant. Unfortunately, some proteins are denatured under the relatively strong oxidizing conditions, and so other methods of radioiodination that employ more gentle conditions have been devised, e.g., the lactoperoxidase method.

The Lactoperoxidase Method

This method, introduced by Marchalonis (*3*), employs lactoperoxidase in the presence of a trace of hydrogen peroxide to oxidize the radioactive iodide. The oxidation can be stopped by simple dilution. Although the technique should result in less chance of denaturation of susceptible proteins than the chloramine-T method, it is more technically demanding and is subject to a more marked variation in optimum reaction conditions.

Materials

The Chloramine-T Method

1. $Na^{125}I$: 1 mCi, concentration 100 mCi/mL.
2. Buffer A: 0.5*M* sodium phosphate buffer, pH 7.4.
3. Buffer B: 0.05*M* sodium phosphate buffer, pH 7.4.
4. Buffer C: 0.01*M* sodium phosphate buffer containing 1*M* sodium chloride, 0.1% bovine serum albumin, and 1% potassium iodide, final pH 7.4.
5. Chloramine-T solution: A 2 mg/mL solution in buffer B is made just prior to use.
6. Reductant: A 1 mg/mL solution of sodium metabisulfite in buffer C is made just prior to use.
7. Protein to be iodinate: A 0.5–2.5 mg/mL solution is made in buffer B.

The Lactoperoxidase Method

1. $Na^{125}I$: 37 MBq (1 mCi) concentration 3.7 GBq/mL (100 mCi/mL).
2. Lactoperoxidase: Available from various commercial sources.
   A stock solution of 10 mg/mL in 0.1*M* sodium acetate buffer, pH 5.6, can be made and stored at −20°C in small aliquots. A working solution of 20 μg/mL is made by dilution in buffer just prior to use.
3. Buffer A: 0.4*M* sodium acetate buffer, pH 5.6.
4. Buffer B: 0.05*M* sodium phosphate buffer, pH 7.4.

5. Hydrogen peroxide: A solution of 10 µg/mL is made by dilution just prior to use.
6. Protein to be iodinated: A 0.5–2.5 mg/mL solution is made in buffer B.

It is essential that none of the solutions contain sodium azide as an antibacterial agent since it inhibits lactoperoxidase.

Method

The Chloramine-T Method

1. Into a small plastic test tube (1 × 5.5 cm) are added successively the protein to be iodinated (10 µg), radioactive iodide (5 µL), buffer A (50 µL), and chloramine-T solution (25 µL).
2. After mixing by gentle shaking, the solution is allowed to stand for 30 s to allow radioiodination to take place.
3. Sodium metabisulfite solution (500 µL) is added to stop the radioiodination and the resultant solution is mixed. It is then ready for purification.

The Lactoperoxidase Method

1. Into a small plastic test tube (1 × 5.5 cm) are added in turn the protein to be iodinated (5 µg), radioactive iodide (5 µL), lactoperoxidase solution (5 µL), and buffer a (45 µL).
2. The reaction is started by the addition of the hydrogen peroxide solution (10 µL) with mixing.
3. The reaction is stopped after 20 min by the addition of buffer B (0.9 mL) with mixing. The resultant solution is then ready for purification.

Purification of Radioiodinated Protein

At the end of the radioiodination the reaction mixture will contain the labeled protein, unlabeled protein, radioiodide, mineral salts, enzyme (in the case of the lactoperoxidase method), and possibly some protein that has been damaged during the oxidation. For most uses of radioiodinated proteins, it is essential to have the labeled species as pure as possible with the constraints, however, that the purification is achieved as rapidly as possible. For that purpose the most widely used of all separation techniques is gel filtration. Various types of Sephadex resin can be employed, e.g., G-50, G-75, and G-100 depending on the differences in sizes of the molecules present in the mixture.

Typically the mixture is applied to a column (1 × 25 cm) of Sephadex resin and is eluted with 0.05*M* sodium phosphate buffer of pH 7.4 containing 0.15*M* sodium chloride and 0.1% bovine serum albumin. Fractions (0.5–1.0 mL) are collected in plastic tubes and aliquots (10 μL) are counted. Using those results, an elution profile such as that shown in Fig. 1 is drawn.

Notes

1. Several parameters can be used to assess the quality of the labeled protein. The *specific radioactivity* of the protein is the amount of radioactivity incorporated per unit mass of protein. It can be calculated in terms of the total radioactivity employed, the amount of the iodination mixture transferred to the gel filtration column, and the amount of radioactivity present in the labeled protein, in the damaged components, and in the residual radioiodine. However, in practice, the calculation does not usually take into account damaged and undamaged protein. The specific activity is thus calculated from the yield of the radioiodination procedure, the amount of radioiodide and the amount of protein used, assuming that there are not significant losses of those two reactants. The *yield of the reaction* is simply the percentage incorporation of the radionuclide into the protein.

   For example, consider the results shown in Fig. 1. In terms of elution volumes, it is to be expected that the first peak of radioactivity represents the labeled protein and that the second peak represents

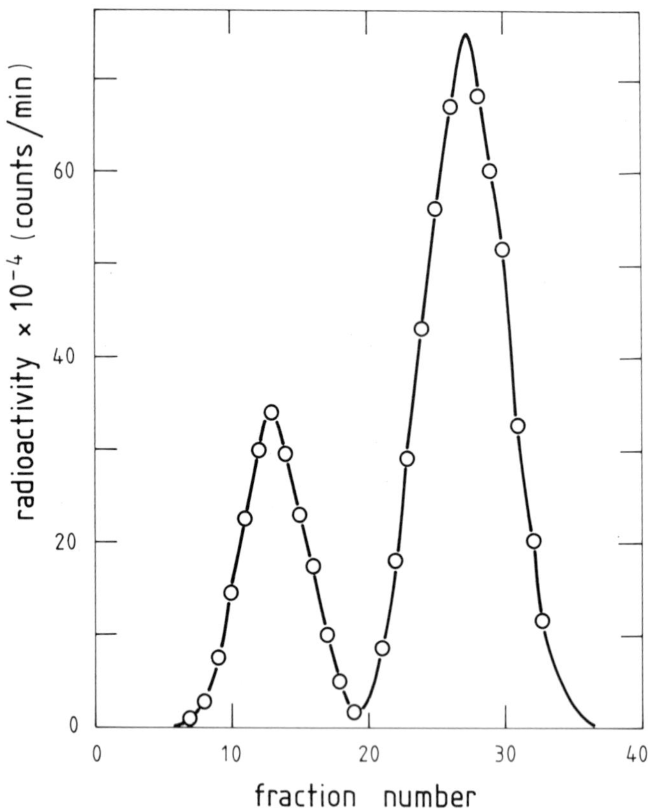

Fig. 1. Gel filtration of radioiodinated kallikrein of rat submandibular gland. The pure enzyme (10 μg) was iodinated with ^{125}I (18.5 MBq) by the chloramine-T method. It was then purified on a column (1 × 20 cm) of Sephadex G-75 resin at a flow rate of 20/mL/h and collecting fractions of 0.6 mL. Aliquots (10 μL) of each fraction were measured for radioactivity. By radioimmunoassay, immunoreactive protein was found only in the first peak and more than 90% of that radioactivity was bound by the antiserum to kallikrein from rat submandibular gland.

unreacted radioiodide. For this case, and, more importantly, for more complicated elution patterns, the nature of the materials giving rise to those peaks can be checked by employing a specific antiserum to the protein being radioiodinated. Aliquots (10 μL) of different fractions making up the two peaks are diluted so that each gives the same number of counts (e.g., 5000–10,000 counts/min) per 100 μL. Those samples are incubated with an excess of the antiserum. Only samples containing immunoreactive protein will react with the antiserum. The amount of radioactive protein associated with the antibody molecules can then be measured by radioimmunoassay (*see* Chapter 37).

Having identified the peak or peaks containing the radioiodinated protein, the yield of the radioiodination can be calculated in terms of the ratio of the total counts associated with the protein peak to the sum of the total counts associated with the protein peak and total counts associated with the iodide peak.

It is obviously important that the radioiodinated protein should as far as possible have the same properties as the unlabeled species. Thus the behavior of both molecules can be checked on electrophoresis or ion-exchange chromatography. The ability of the two species to bind to specific antibodies can be assessed by radioimmunoassay.

2. To store the labelled protein, immediately after purification split the sample into small aliquots and then rapidly freeze and store at −20°C. Alternatively the aliquots can be freeze-dried. Each aliquot should be melted and used only once. Radioiodinated proteins differ markedly in their stability. Some can be stored for several weeks (though it must be borne in mind that the half-life of ^{125}I is about 60 d), whereas others can only be kept for several days. If necessary, the labeled protein can be repurified by gel filtration or ion-exchange chromatography prior to use.
3. The pH optimum for iodination of tyrosine residues of a protein by the chloramine-T method is about pH 7.4. Lower yields of iodinated protein are obtained at pH values below about 6.5 and above about 8.5. Indeed, above pH 8.5, the iodination of histidine residues appears to be favored.
4. The total volume of the chloramine-T reaction mix should be as low as practically possible to achieve a rapid and efficient incorporation of the radioactive iodine into the protein. Because of the small volumes of reactants that are employed it is essential to ensure

adequate mixing at the outset of the reaction. Inadequate mixing is one of the commonest reasons for a poor yield of radioiodinated protein by this procedure.

5. If the protein has been seriously damaged by the use of 50 μg of chloramine-T, it may be worthwhile repeating the radioiodination using much less oxidant (10 μg or less). Obviously the minimum amount of chloramine-T that can be used will depend, among other factors, on the nature and amount of protein to be iodinated.
6. It is normal to carry out the chloramine-T method at room temperature. However, if the protein is especially labile, it may be beneficial to run the procedure at a low temperature.
7. For the lactoperoxidase method, the exact nature of buffer A will depend on the properties of the protein to be radioiodinated. Proteins differ markedly in their pH optima for radioiodination by this method (*4*). Obviously the pH to be used will also depend on the stability of the protein and the optimum pH can be established by trial and error.
8. Other reaction conditions such as amount of lactoperoxidase, amount and frequency of addition of hydrogen peroxide, and so on also markedly affect the yield and quality of the radioiodinated protein. Optimum conditions can be found by trial and error.
9. The longer the time of the incubation the greater is the risk of potential damage to the protein by the radioactive iodide. Thus it is best to keep the time of exposure of the protein to the radioactive iodide as short as possible, but commensurate with a good yield of radioactive product.
10. Some of the lactoperoxidase itself may become radioiodinated, which may result in difficulties in purification if the enzyme is of a similar size to the protein being labeled. Thus it is best to keep the ratio of the amount of protein being labeled to the amount of lactoperoxidase used as high as possible.
11. Some of the problems may be overcome by the use of solid-phase lactoperoxidase systems. Such a system is commercially available in which immobilized glucose

oxidase is used to generate a small, steady stream of hydrogen peroxide from added glucose. The hydrogen peroxide is utilized by immobilized lactoperoxidase to oxidize the radioactive iodide.

References

1. Bolton, A. E. (1977) "Radioiodination Techniques" Amersham International, Amersham, Bucks, England.
2. Hunter, W. M., and Greenwood, F. C. (1962) Preparation of iodine-131 labeled human growth hormone of high specific activity. *Nature* **194,** 495–496.
3. Marchalonis, J. J. (1969) An enzymic method for trace iodination of immunoglubulins and other proteins. *Biochem. J.* **113,** 299–305.
4. Morrison, M., and Bayse, G. S. (1970) Catalysis of iodination by lactoperoxidase. *Biochemistry* **9,** 2995–3000.

Chapter 37

Radioimmunoassay

G. S. Bailey

Department of Chemistry, University of Essex, Colchester, Essex, England

Introduction

Radioimmunoassay is often described in terms of the competition between a radiolabeled antigen (Ag^*) and its unlabeled counterpart (Ag) for binding to a limited amount of specific antibody (Ab) (*1*). In most radioimmunoassays the reaction is allowed to proceed to equilibrium and thus, can be represented by Eq. (1).

$$Ag^* + Ag + 2Ab \rightleftharpoons Ag^*Ab + AgAb \quad (1)$$

The concentration of the antibody is limited such that the labeled antigen, although present in a trace amount, is in relative excess over the antibody. Thus, even in the absence of unlabeled antigen, only some of the radioactive antigen will be associated with the antigen–antibody complex while the remainder will be free in solution. In the radioimmunoassay, the total amounts of antibody and radiolabeled antigen are kept constant. The presence of unlabeled antigen will result in less of the labeled species being able to bind to the antibody. The greater the amount of unlabeled antigen (Ag) present, the lower will be the amount of radiolabeled antigen combined to the antibody (Ag*Ab). Thus, on suitable calibration, the amount of the unlabeled species can be accurately measured in terms of the amount of radioactivity associated with the antigen–antibody complex.

The widespread use of radiommunoassays results from several significant advantages that the method has compared to many other quantitative assays, particularly for substances that are otherwise measured by pharmacological assays. Those advantages include very high sensitivity and high specificity. A wide variety of exogenous and endogenous substances can be measured in biological tissues and fluids at concentrations of pg/mL by the use of suitable radioimmunoassays. Such high sensitivity requires the use of antisera of high avidity and the use as tracers of antigens labeled to high specific radioactivities. The production of antisera and the radioiodination of proteins are dealt with in other chapters.

1. Incubation of Antigens and Antiserum. The incubation conditions have to ensure the stability of all reagents as antigen binds to antibody and allow equilibrium to be reached.
2. Separation of Free and Bound Antigen. When equilibrium has been achieved, the antigen bound to antibody is quickly and efficiently separated from free antigen so that the radioactivity associated with either or both components can be counted. Many different separation procedures have been reported (2). This chapter will deal with two which are widely used.

Fractional Precipitation with Polyethyleneglycol

The basis of the method is the ability of relatively low concentrations of polyethyleneglycol to bring about the precipitation of antibody molecules, presumably by removal of the attendant hydration shell of water molecules, without the precipitation of the smaller antigen molecules (3). The method is not efficient in all cases, but is worth trying because of its simplicity and very low cost.

Double (Second) Antibody Method

This procedure is very widely used and can achieve an efficient separation of free and bound antigen in more or less all radioimmunoassays. The basis of the most common type of this method is to use an antiserum (the second antibody), raised to antibodies of the antiserum (first antibody) employed in the incubation, to precipitate the antigen–antibody complex (*4*). The addition of non-immune γ-globulin of the species in which the first antiserum was raised increases the bulk of material that can interact with the second antiserum and so enables a precipitate to be formed.

Materials

Incubation of Antigens and Antiserum

1. Antiserum. The antiserum to be used in the assay should be of high avidity, high titer, and high specificity for the antigen to be measured.
2. Radiolabeled Antigen. Pure antigen must be used for labeling. Purification of the labeled species must also be carried out. Radioiodination is the method of choice for labeling proteins (*see* Chapter 36) to high specific radioactivity.
3. Unlabeled Antigen. Pure antigen is used as the standard in the assay. It is essential that the standard and the antigen to be measured show identical behavior towards the antiserum.
4. Buffer. Many different buffers can be used for radioimmunoassay. Whatever buffer is chosen must ensure the stability of all of the reagents. In practice, in order to achieve equilibrium in a reasonable time, as well as maintain stability, the buffer employed has a pH within the range pH 6–8.6 and a molarity within the range 0.01–0.1*M*. A preservative such as sodium azide or sodium ethyl mercurithiosalicylate (0.01–0.1%) has to be included. Proteins such as bovine serum albumin

or ovalbumin (0.1–1.0%) are normally added as a carrier.

5. Disposable Plastic Tubes. Tubes of various shapes and sizes can be used, depending on the volume of the incubation mixture and the separation procedure employed.

Separation of Free and Bound Antigen

Fractional Precipitation Using Polyethyleneglycol

1. Polyethyleneglycol of molecular weight 6000.
2. Bovine γ-globulin (Cohn Fraction II).

Double (Second) Antibody Method

1. Antiserum to γ-globulins of the species in which the first antiserum (i.e., that used in the incubation) was raised. Such antisera are available from several commercial sources.
2. Non-immune γ-globulin (Cohn Fraction II) or serum of the species in which the first antiserum was raised.

Method

Incubation of Antigens and Antiserum

1. At the outset, the amount of labeled antigen to be employed in the incubation is chosen to be of the same order of magnitude as the smallest amount of unlabeled antigen that is required to be measured. In practice the amount of tracer used is often simply that which gives a specified number of counts per min per unit volume (usually 5000–10,000 counts/min/100 μL).
2. The optimal dilution (titer) of the antiserum to be used in the assay is that which will bind 30–50% of the labeled antigen. That dilution is chosen with the aid of an antiserum dilution curve. The curve is constructed from the results of incubations, carried out at least in

duplicate, of serial dilutions of the antiserum with the chosen amount of tracer. The incubations, for example, consisting of diluted antiserum (100 μL), tracer (100 μL) and buffer (200 μL), are allowed to proceed to equilibrium. The time required to reach equilibrium obviously depends on the particular circumstances of the assay. For a completely new radioimmunoassay, it may be necessary to run pilot experiments varying temperature from 4 to 37°C and time of incubation from 4 to 72 h to establish the optimal conditions. The bound and free forms of the radioactive labeled tracer are then separated and counted. A typical antiserum dilution curve is shown in Fig. 1. In that particular case, the dilution of antiserum to be used in the assay was chosen as 1:60,000 (i.e., final dilution 1:240,000), corresponding to 37% binding of the tracer. The slope of the dilution curve in its descending portion can be

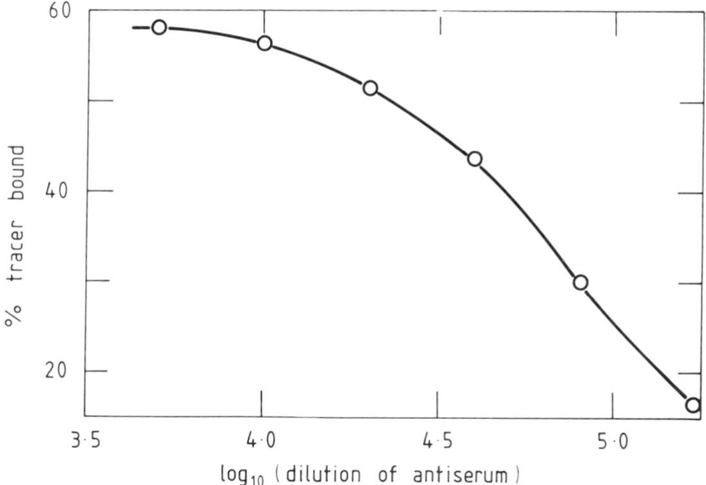

Fig. 1. Antiserum Dilution Curve. Radioiodinated kallikrein (100 μL; 7000 counts/min) was incubated with serial dilutions (1:5000–1:160,000) of antiserum (100 μL) to kallikrein from rat submandibular gland in a total volume of 400 μL for 20 h at 25°C. The bound antigen was separated from the free antigen by incubation at 37°C for 2 h with a solid-phase second antibody. The bound radioactivity was measured and was expressed as a percentage of the total radioactivity that was used.

used as a measure of the avidity of the antiserum (the greater the slope, the greater is the avidity).

3. Having chosen the amount of tracer and dilution of antiserum to be used, the radioimmunoassay can then be set up to measure the amount of antigen in unknown solutions with the aid of standard solutions of antigen.
4. Standard solutions of antigen are made by dilution of a master solution of accurately known concentration. The working solutions are made just prior to use and are kept at 4°C until required.
5. The tubes to be used in the incubation are numbered.
6. Buffer is added to all of the tubes apart from the first three tubes, which will be used to measure the total counts in the assay. To each of those three tubes that will be used to measure nonspecific binding is added 300 μL of buffer. To each of the three tubes that will represent the zero standard is added 200 μL of buffer, while 100 μL of buffer is added to every other tube. The standard solutions (100 μL) of antigen or unknown solutions are then added to the relevant numbered tubes in duplicate. Next the antiserum (100 μL) at the chosen dilution is added to all tubes except for the three tubes to be used to measure total counts and the three tubes to be used to measure nonspecific binding. Finally the radioactive tracer (100 μL) is added to every tube. The contents of each tube are thoroughly mixed with the aid of a vortex mixer. Each tube is then left at a constant temperature until equilibrium is reached (often 24–48 h at 4°C).
7. When equilibrium has been reached the free and bound antigen are separated.
8. Although it is possible for either or both phases to be counted, it is normal for the radioactivity associated with the antibody to be measured. There should be good agreement between the counts of each tube in a particular pair or set of three.
9. The counts associated with the three tubes representing nonspecific binding are averaged and in turn are subtracted from the average counts for each set of tubes.

10. The resultant, average specific counts associated with the three zero-standard tubes can be expressed as a percentage of the average specific counts of the three tubes containing just tracer. If the conditions of the assay have been satisfactory, the counts of the zero standard should be 30–50% of the total counts of the tracer.
11. If so, the average specific counts of each set of duplicates can then be expressed as a percentage of the average specific counts of the three zero standards.
12. A standard curve is constructed from the calculated results.

Separation of Free and Bound Antigen

Fractional Precipitation Using Polyethylene Glycol

1. A solution (25–30%) of polyethylene glycol 6000 is prepared in 0.05*M* sodium phosphate buffer, pH 7.4, with thorough mixing. A separate solution (1%) of bovine γ-globulin is made in the same buffer.
2. To each of the tubes requiring separation of bound and free antigen is added at 4°C γ-globulin solution (400 μL) and polyethylene glycol solution (800 μL). Each tube is vortexed and is allowed to remain at 4°C for 15 min.
3. Each tube is then centrifuged at 4°C at 5000*g* for 30 min.
4. The supernatants are carefully removed by aspiration at a water pump. The precipitate at the bottom of each tube is then counted. If necessary, the precipitates can be washed at 4°C with 15% polyethylene glycol solution (800 μL) followed by vortexing and centrifugation.

Double (Second) Antibody Method

1. Firstly, the optimal dilution of the second antiserum has to be carefully assessed. Serial dilutions of the second antiserum are made in 0.05*M* sodium phosphate buffer, pH 7.4, containing 0.5% non-immune serum and 50 m*M* EDTA.

2. Samples containing tracer (100 μL), first antiserum at optimal dilution (100 μL), and buffer (200 μL) are allowed to come to equilibrium at constant temperature.
3. Aliquots (100 μL) of the different dilutions of the second antiserum are added. The tubes are vortexed and are kept at constant temperature for a further 18–24 h. The tubes are then centrifuged at 5000*g* for 30 min. The supernatants are aspirated at a water pump and the tubes are counted.
4. The highest dilution of the second antiserum that results in the precipitation of a maximum percentage of the tracer represents the minimum amount of second antiserum that should be employed in the separation.
5. Aliquots (100 μL) of the chosen dilution of the second antiserum are added to each of the sample tubes requiring separation of free and bound antigen. The tubes are then treated as in point 3 above.

Notes

1. The purpose of a radioimmunoassay is to measure the concentration of a particular antigen in an unknown sample by comparison with standard solutions of the antigen. From the measurements of the bound or free radioactivity in the presence of various known amounts of antigen is constructed a standard curve. The amount of antigen in the test sample is then found from that curve by simple interpolation. The standard curve can be represented in several different ways. One straightforward method is to plot the proportion of tracer bound expressed as a percentage of that in the zero standard against the corresponding concentration of standard antigen (*see* Fig. 2). However, to avoid the problems associated with the subjectivity of drawing a curve, it is better to construct a linear plot. One such linear transformation that is very widely used is shown in Fig. 3, where the logit of the tracer bound (y) is plotted against log concentration.

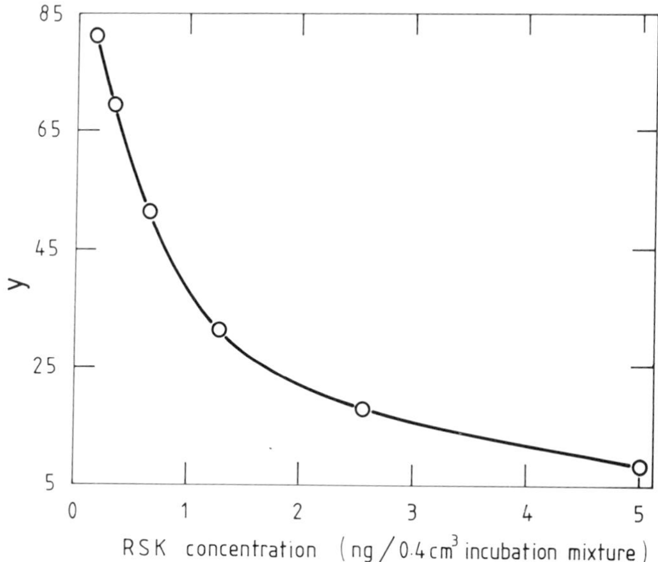

Fig. 2. Standard Curve for Radioimmunoassay. Radioiodinated kallikrein (100 μL; 7000 counts/min) was incubated with standard solutions of kallikrein from rat submandibular gland (200 μL containing 0.16–5.12 ng kallikrein) and antiserum to the enzyme (100 μL of 1:60,000 dilution) for 20 h at 25°C. The bound radioactivity was separated from the free radioactivity by use of a solid-phase second antibody on incubation from 2 h at 37°C and was counted.

$$y = \frac{B - N}{B_o - N} \times 100$$

where

B = counts associated with standard solution
B_o = counts associated with zero standard
N = counts associated with nonspecific binding.

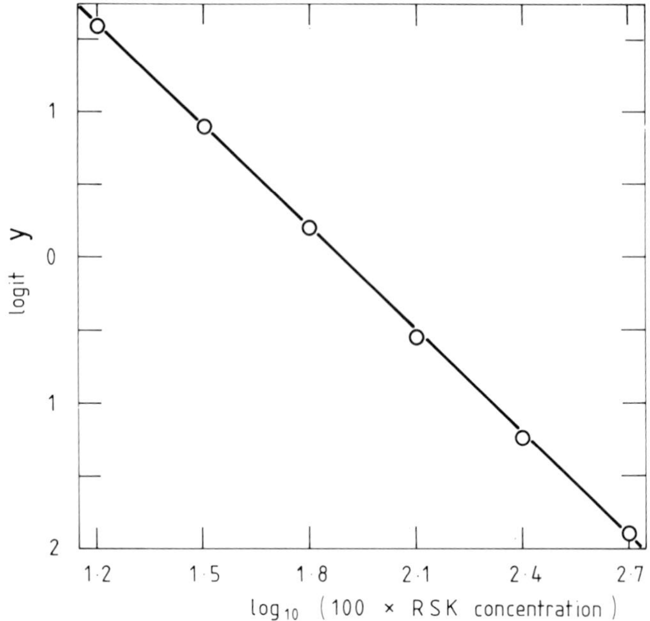

Fig. 3. Logit-log plot of the standard curve. For details see legend to Fig. 2.

$$\text{logit } y = \log_e \left(\frac{y}{100 - y} \right)$$

and

$$y = \frac{B - N}{B_o - N} \times 100$$

where B = counts associated with a certain concentration of antigen
B_o = counts associated with the zero standard
N = counts associated with nonspecific binding.

Often the logit plot becomes nonlinear near its extremes and in that case the points in those regions are not employed in construction of the straight line.

2. There are a very large number of variations to the procedure outlined. The best set of conditions for a particular case can be worked out by trial and error. Of particular importance is the volume of the unlabeled

antigen in the incubation mixture. In theory, an increase in that volume relative to the other components should lead to an increase in sensitivity of the assay, i.e., detection of a lower amount of antigen. In practice, there may actually be a decrease in sensitivity because of interference in the antigen–antibody binding by other substances present in the sample.

3. Other factors that often increase the sensitivity of the assay include
   (a) 'Late' addition of the tracer. The labeled antigen is added a considerable time after the unlabeled species and antiserum have been allowed to interact, but before the attainment of equilibrium.
   (b) Decrease in the amount of antiserum. However, a limit will be reached beyond which a further reduction in antiserum will be counter-productive because of a loss of precision in the assay.
   (c) Decrease in the amount of tracer. Again, there is a limit beyond which a further reduction will produce no significant change in sensitivity.

   It should be noted, however, that very high sensitivity may not be required of a radioimmunoassay. What is far more important is that the assay should be able to accurately, precisely and reproducibly measure antigen in the range of concentrations found in biological tissues and fluids, preferably without the need for sample dilutions. However, the following note should be borne in mind.

4. One major problem of radioimmunoassay, particularly when applied to undiluted biological samples, is that substances in the samples interfere with antigen–antibody interaction. High concentrations of salts and plasma proteins decrease antigen–antibody binding. Thus it may be necessary to dilute the samples to reduce such effects or to include such substances in the standard antigen solutions.
5. Another problem may be encountered if the biological samples contain proteolytic enzymes that will degrade the antigen. In those cases, it is necessary to include a broad-spectrum inhibitor in the assay, e.g., aprotinin.

6. Precipitation by polyethylene glycol is very sensitive to fluctuations in temperature. Hence it is essential to keep all solutions and tubes at 4°C during the whole procedure.
7. One problem that is often associated with the method of fractionation using polyethylene glycol is the occurrence of significant nonspecific binding, i.e., high blanks. Washing of the precipitate with polyethylene glycol solution may reduce the nonspecific binding.
8. When carrying out the double (second) antibody method, normal serum or non-immune γ-globulin from the species in which the first antiserum was raised can be included in the incubation mixture rather than be added with the second antiserum.

   However, the presence of components of the complement system in serum samples can affect the formation of the immune precipitate. Thus it is usual to include EDTA (0.05–0.1*M*) in the incubation and separation buffers to inactivate those components and avoid any such difficulties.
9. If the concentrations of reactants are low, it can take a considerable time for complete immunoprecipitation to occur. The use of higher amounts of γ-globulin or serum will greatly speed up the process, but of course higher amounts of the second antiserum will also be needed.
10. Many of the difficulties associated with the double antibody method can be avoided through the use of a second antibody that is covalently linked to a suitable, inert, solid matrix. Because of the solid-phase nature of such a second antibody, immune precipitation occurs relatively quickly and a carrier non-immune γ-globulin is not required. Of course, considerable effort, second antiserum, and a suitable solid support are necessary to produce the solid-phase second antiserum (*5*). Certain preparations are commercially available, but are expensive, e.g., antibodies raised in sheep against rabbit γ-globulins, coupled to cellulose; antibodies raised in goat against rabbit γ-globulins and coupled to polyacrylamide.
11. One particularly useful modification of the double antibody method is to include a small amount of ammo-

nium sulfate or polyethylene glycol with the second antiserum (*6*). The inclusion of the chemical precipitant speeds up the immune precipitation and enables a lower amount of second antiserum to be used.

References

1. Felber, J. P. (1975) Radioimmunoassay of Polypeptide Hormones and Enzymes in *Methods of Biochemical Analysis 22* (ed. Glick, D.) pp. 1–94. Wiley, New York.
2. Chard, T. (1978) Requirements for a binding assay—separation of bound and free ligand, in *Laboratory Techniques in Biochemistry and Molecular Biology* **6** (eds., Work, T. S., and Work, E.) pp. 401–426. Elsevier/North Holland Biomedical Press, Amsterdam.
3. Desbuquois, B., and Aurbach, G. D. (1971) Use of Polyethylene Glycol to Separate Free and Antibody-Bound Peptide Hormones in Radioimmunoassay. *J. Clin. Endocrinol. Metab.* **33,** 732–738.
4. Midgley, A. R., and Hepburn, M. R. (1980) Use of the double-antibody method to separate antibody bound from free ligand in radioimmunoassay in *Methods in Enzymology* **70** (eds. Van Vunakis, H., and Langone, J. J.) pp. 266–274. Academic Press, New York.
5. Koninckx, P., Bouillon, R., and De Moor, P. (1976) Second antibody chemically linked to cellulose for the separation of bound and free hormone: An improvement over soluble second antibody in gonadotrophin radioimmunoassay. *Acta Endocrinol.* **81,** 43–53.
6. Peterson, M. A., and Swerdloff, R. S. (1979) Separation of bound from free hormone in radioimmunoassay of lutropin and follitropin. *Clin. Chem.* **25,** 1239–1241.

Chapter 38

Enzyme-Linked Immunosorbant Assay (ELISA)

Wim Gaastra

Department of Microbiology, The Technical University of Denmark, Lyngby, Denmark

Introduction

In general, immunological methods are not very well suited for a quantitative determination of the antigen to be studied. The ELISA technique, however, can be used for a quantitative or at least semiquantitative determination of the concentration of a certain antigen. The method was first introduced by Engvall and Perlmann (*1*). The principle of ELISA (see Fig. 1), also called the double antibody sandwich technique, is the following: Antibodies against the antigen to be measured are adsorbed to a solid support, in most cases a polystyrene microtiter plate. After coating the support with antibody and washing, the antigen is added and will bind to the adsorbed antibodies.

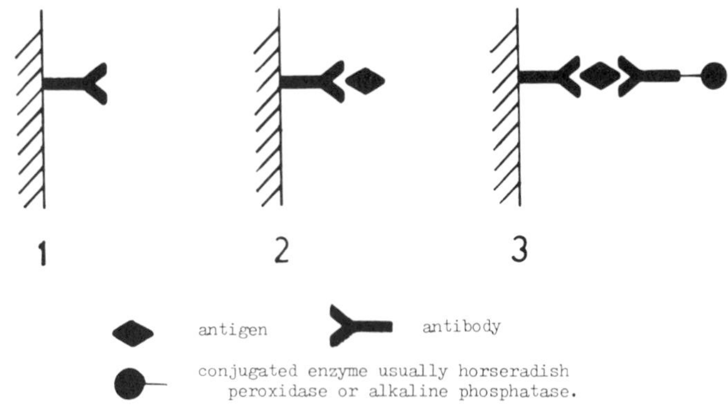

Fig. 1. The main steps in a (noncompetitive) ELISA test. (1) The antibody to the antigen being quantitated is adsorbed onto a solid phase, usually polystyrene. (2) The sample containing the antigen being measured is then added. (3) Following the incubation and washing steps, a second enzyme-labeled antibody is then added. After further incubation and washing steps, enzyme substrate is added. (A substrate is chosen that will give a colored product). The amount of color produced is therefore proportional to the amount of antigen bound to the original antibody.

Next, a conjugate that will also bind to the antigen is added. Conjugates are antibody molecules to which an enzyme is covalently bound.

After addition of a chromogenic substrate for the enzyme, the intensity of the colored reaction products generated will be proportional to the amount of conjugated enzyme and thus indirectly to the amount of bound antigen molecules. Since the intensity of the developed color is proportional to the amount of antigen molecules present, determination of the intensity of the color produced by a standard series of antigens will allow the calculation of the amount of antigen in an unknown sample. This chapter describes the double antibody sandwich technique together with a method for preparing conjugates.

Materials

1. Microtiter plates, 96 wells, flat-bottomed polystyrene.
2. Multichannel pipet: 0–250 μL (8-channel), used for pipeting of all reagents.

3. Gilson or Finnpipettes: ranging from 0 to 250 μL used for pipeting of blanks, standards, and samples into wells.
4. Titertek multiscanphotometer or equivalent if available.
5. 0.1*M* carbonate buffer (pH 9.6). (0.1*M* Na_2CO_3 brought to pH 9.6 with NaOH.) This buffer should be made of the purest chemicals available and double-distilled H_2O.
6. Wash solution. 90 mL Tween 80 added to 910 mL H_2O.
7. BST: 0.2% (w/v) BSA (bovine serum albumin), 0.01% Tween 80 and 0.9% (w/v) NaCl in distilled water.
8. Substrate solutions. Depending on which enzyme is coupled to the conjugate (*see* Notes), different substrate solutions have to be used. The substrate solutions given here have to be used when horseradish peroxidase (HRP) is the enzyme coupled to the conjugate. There are two substrate solutions in use for HRP. One containing 5-amino-salicylic acid (purple-redbrown color) and one containing ortho-phenylene diamine (yellow color). Both compounds are light-sensitive and the solutions containing them must be freshly made and protected against light. Both substrate solutions are possibly carcinogenic.

   Solution 1: Dissolve 80 mg 5-amino salicylic acid in 100 mL 0.05*M* potassium phosphate buffer (pH 6.0) containing 0.001*M* EDTA. Add 20 mL H_2O_2 (30%) and mix.

   Solution 2: Mix: 24.3 mL 0.1*M* citric acid
25.7 mL 0.2*M* Na_2HPO_4
50 mL H_2O
40 mg ortho-phenylene diamine
40 μL H_2O_2 (30%)

   In the case of solution 1, 0.3*M* NaOH is used to stop the reaction; when solution 2 is used the reaction is stopped with 1*M* H_2SO_4.

9. A diluted solution of IgG against the antigen to be measured. Depending on the quality of the IgG (titer), the dilution is usually 1500–2500-fold. Dilutions are made in 0.1*M* carbonate buffer.

10. Antigen solutions to be tested and standard antigen solutions.
11. HRP-labeled diluted IgG solution. As a rule the conjugate solution has to be diluted 500–2000 times in BST (but *see also* Notes section).

Method

The Double Antibody Sandwich Technique

1. Add 150 µL of the diluted IgG solution to each of the wells of a microtiter plate (or plates) using the 8-channel Titertek pipet. Cover the plate(s) and incubate for 16 h at room temperature.
2. Wash the plate(s). The wells are emptied by flicking the plate over a sink. Residual liquid is removed by "beating" the plate upside-down against a filterpaper. If kept covered and cool, coated plates can be kept for a substantial period of time, up to several months). An appropriate amount of wash solution is pipetted into the wells and left for a couple of minutes. The wells are then emptied as described above. This procedure is then repeated once. If a special washing device is used (i.e., a Titertek Microplate Washer or a homemade device that can be connected to a waterpump), the plates are washed three times for 15 s with the washing solution, after having been emptied each time as described above. Good washing is essential. One should rather wash for 20 than 10 s.
3. After washing add 100 µL BST to every well, using the multichannel pipet.
4. Add 100 µL of an antigen solution to be tested to the first well of each row. Mix carefully by taking the solution up and down with the multichannel pipet several times. Care should be taken to avoid air bubbles or splashing of small drops.
5. Take 100 µL from the first wells and transfer them to the second wells with the multichannel pipet. Repeat the mixing procedure. Take 100 µL from the second wells and add to the third and so on. In this way a

two-fold dilution series from wells 1–12 is created. Finally, remove the 100 μL excess from the last wells.

6. Incubate the plate for 2 h at 37°C, to let the antigen bind to the coated antiserum, and then wash thoroughly.
7. Add 100 μL of the diluted conjugate solution to all wells and incubate for 2 h at 37°C, then wash the plate thoroughly.
8. Add 100 μL of substrate solution to all wells and incubate for 1–2 h at 37°C in the dark.
9. Stop the reaction with 100 μL of stop solution.
10. Read the titer of the antigen solutions. This can be done by either using a Titertek multichannel photometer or by determining the last well that still gives some coloring observable with the naked eye. The incubations with the antigen and conjugate solutions should be done while the plates are continuously gently shaken. This is not absolutely necessary for the incubation with the substrate solution.

Preparation of a Conjugate for ELISA

As stated above, a conjugate is the covalent complex of IgG and an enzyme. In the procedure below, HRP is the enzyme that is coupled. See Notes section for other enzymes that can be used.

1. Dissolve 5 mg of horseradish peroxidase in 1 mL 0.3*M* Na_2CO_3 (pH 8.1). This solution should be prepared fresh.
2. Add 0.1 mL of 1% fluorodinitrobenzene in pure ethanol. (**N.B.** Fluorodinitrobenzene is suspected to be carcinogenic.) If the HRP used is not pure, a precipitate may be formed that must be removed by centrifugation (10 min, 18,000 rpm).
3. Mix thoroughly and incubate for 1 h at room temperature.
4. Add 1 mL of 0.16*M* ethylene glycol, mix, and incubate for another hour at room temperature. The total volume is now 2.1 mL.

5. Dialyze the mixture against 0.01*M* sodium carbonate buffer (pH 9.5) for 25 h. The carbonate buffer should be changed at least three times.
6. Add IgG (*see* below) dissolved in 0.01*M* sodium carbonate buffer (pH 9.5) to the peroxidase-aldehyde solution in the following ratio: one volume IgG solution to one volume activated peroxidase-aldehyde, or 5 mg purified IgG (protein) to 3 mL peroxidase solution.
7. Mix well and incubate 2–3 h, but not longer at room temperature. In the case of precipitate formation, the precipitate should be removed by centrifugation (10 min, 10,000 rpm).
8. Dialyze extensively against 0.01*M* phosphate buffer (pH 7.2) containing 0.9% NaCl at 4°C. Store the conjugate in a refrigerator or freezer.
   N.B. Repeated freezing and thawing of IgG and conjugate solutions diminishes the activity of the proteins. Thawed solutions that are not completely used can be very well kept at 4°C.

Preparation of a Purified IgG Fraction from Whole Serum

1. Take 100 mL serum and add 200 mL of 0.06*M* sodium acetate (pH 4.6). The pH of the mixture should be 4.8.
2. Add 8.2 mL caprylic acid dropwise at room temperature (but *see also* Notes section).
3. Stir for 30 min and remove the precipitate (10 min, 10,000 rpm).
4. Dialyze the IgG fraction against 0.9% sodium chloride and lyophilize.

Notes

1. The amount of caprylic acid needed for purification of an IgG fraction varies from sera to sera as indicated in this table.

100 mL serum	Caprylic acid, mL
Rabbit	8.2
Horse	7.6
Human	7.6
Bovine	6.8
Goat	8.0

2. The procedure described above can be varied in many ways. For example, one can add a fixed amount of enzyme labeled antigen to the sample. This will then compete with the unlabeled antigen and from the degree of competition in regard to a standard antigen solution, the concentration of the antigen can be evaluated. In some cases, the antigen can be coated to the wall of the polystyrene plate and then incubated with the conjugate directly. In these cases then, there is no sandwich of the antigen. The principle of the method however, remains the same.
3. Alkaline phosphatase is the other enzyme used in ELISA tests to prepare conjugates. In this case the substrate is *p*-nitrophenylphosphate that the enzyme converts to the yellow *p*-nitrophenol that is measured at 405 nm.
4. Occasionally an ELISA test will fail to give the desired result and in all wells the same amount of colour will develop. In most cases this problem can be overcome by preparing fresh solutions.

References

1. Engvall, E., and Perlmann, P. (1971) *Immunochemistry* **8**, 871–874.

Index